OFF GRID
FARM ANIMAL
ESSENTIALS

OFF GRID FARM ANIMAL ESSENTIALS

A Comprehensive Handbook for Raising Cows, Goats, Sheep, Pigs, Rabbits, Chickens, Ducks, Geese, and Turkeys

Self-Sustaining Solutions for Off Grid Homesteading!

JOHN UTTERBACK

Published by: JEU Publishing

Interior Design: Creative Publishing Book Design

ISBN eBook: 979-8-9875542-3-4
ISBN Paperback: 979-8-9875542-4-1
ISBN Hardback: 979-8-9875542-5-8

Contents

Introduction

You're wanting to be self-sufficient on your homestead, but getting there effectively and efficiently is the challenge. There are many steps and details to consider, and this book can help you along every step of the way in this important journey.

You don't want to buy a bunch of animals and assume that it will all work out somehow. That is a mistake that many make, but with careful planning, you can avoid all of the discouraging losses and unnecessary expenses.

The best approach is to first determine what your goals are for milk, meat, eggs, and fiber. Once that is known, you can choose which farm animals would best meet these needs, and then which breeds are the best fit for your homestead space.

Once you determine which animals you want, and which breeds are best, then you need to think about numbers—how many are required of each type and breed? You should take into account both your necessities and the numbers your homestead can support.

Third, you will need a plan on how to begin—where to purchase your stock, how to house the animals, and what you will need for containment, shelter, and predator protection.

Fourth, a viable breeding regimen for livestock, and a hatching and brooding protocol for poultry will be essential. You will need to be well acquainted with all of the nuances of how to raise and care for each type of animal.

For instance, it will be vital to know how much to feed them and with what kind of feed. It will also be important to understand how to water them, and what kind of equipment will work best. You will need to be familiar with what the common health issues can be for your animals and the precautionary measures necessary to avoid them.

Next, you should be aware of how to manage your products—how to collect, process, and store the products, and how to sell your abundance.

Finally, you should have a way of formulating a cost analysis and a benefit/value assessment, so you can easily and carefully manage expenses.

These are the types of considerations you need to weigh and plan for before you buy your first farm animal. This book will help. It is designed to prepare you for everything you will need to know, so that you can make the right choices and begin with confidence.

The best way to have success in your farm animal experience is to have a good plan and the knowledge required to proceed effectively and efficiently. This book can help you accomplish your goals and more, so that your journey can be fun, and be a memory-making experience for you and your family.

PART 1

Livestock

Part 1 will cover cows, goats, sheep, pigs, and rabbits and the various breeds that are most commonly used to accomplish the goals for milk, meat, and fiber. Some livestock have additional side benefits, such as brush control.

Each chapter will cover management techniques for breeding, feeding, watering, and pasture management. Important requirements for containment and shelter are also discussed. Key elements, such as milk production, meat production, and fiber processing, will be explained in detail.

Each chapter lists common health issues and the importance of health management. At the end of each chapter a cost analysis, and a benefit/value assessment, or a profit/loss analysis will be provided. Finally, each chapter will conclude with a picture gallery of images for the referenced breeds.

Part 1 is full of vital information to help the reader fully understand what is involved in raising livestock in order to be self-sufficient. You will discover which animals will best meet your needs and which breeds will be the best fit for your homestead.

CHAPTER 1

Milk Cows

Having a milk cow on a homestead can be a fun experience for you and your family. However, the decision comes with a set of responsibilities that need to be carefully considered. For instance, a milk cow will require more space and consume more feed than other farm animals and must be milked twice a day (every day).

Other things warranting consideration include annual breeding, an adequate feed source, and a plan for daily milk usage, storage, or processing. Good candidates for a milk cow are homesteads that have appropriate acreage and a means to utilize high volumes of milk.

This chapter will discuss how to choose the best milk cow breed for your needs, as well as the basic requirements for containment, shelter, daily feed costs, daily labor commitment, pasture assessment, animal management, and milk management. This chapter will conclude with a cost/benefit analysis to assist you in the decision-making process.

Dairy Breeds

There are several dairy cattle breeds used for milk production. The following breeds are the most common dairy breeds in the United States.

Holsteins

The Holstein originates from Holland in the Netherlands, and they are best adapted to mild climates. This breed is the most common dairy breed in the US, and they are easily recognized by their black and white markings. Holsteins are known for their ability to produce large quantities of milk but with a lower butterfat content compared to other breeds.

Jerseys

The Jersey breed originates from the Channel Island of Jersey, which is located in the English Channel off the coast of France. This breed is smaller in size compared to Holsteins, but they are good milk producers. They have a dark brown coat and are known for their high butterfat content, which makes their milk ideal for cheese and butter production. Jerseys have a docile temperament and are well suited for grazing. They thrive in pasture-based systems.

Guernseys

This breed originates from the Channel Island of Guernsey, which is also located in the English Channel. They have blond and white markings. They are medium-sized cows and known for their docile temperament and milk that has a high butterfat content. The milk is a golden color and high in beta carotene—an excellent source of vitamin A. Guernseys are well-suited to grazing and are valued for their ability to convert forage into milk efficiently.

Brown Swiss

The Brown Swiss originates from Switzerland, and they are large, sturdy cattle with a distinctive gray/brown color. They are known for their exceptional milk quality, including high protein and butterfat content, and a close fat-to-protein ratio desirable for making cheese. Brown Swiss cattle are well-suited to rugged terrains and can tolerate a range of climates.

Ayrshires

Ayrshires originate from Scotland and are a hardy breed known for their ability to adapt to various climates. They have a red and white coat, and their milk is of good quality, with a moderate butterfat content. Ayrshires are valued for their overall health and longevity.

Other Breeds

While the above-mentioned breeds are the most common in the US, there are other breeds that are used by some dairies, though they are less prevalent. These include the Milking Shorthorn and the Red and White Holsteins. Breed choice is usually based on factors such as milk production, milk quality, adaptability to local conditions, and personal preferences.

Breed Selection Considerations

When choosing a dairy cow, several factors are generally considered to ensure that your specific needs will be best served.

Here are some of the most common considerations.

Milk Production

The quantity of milk produced is often one of the more common considerations. If the daily volume of milk is a priority for you, the Holstein would be the best choice.

Milk Composition

Butterfat content and protein content are also desirable criteria for consideration. The Jerseys and Guernseys are known for their higher butterfat content, which is desirable for dairy products like butter, cheese, and yogurt.

Adaptability

Dairy cows should be well-suited to the local climate, environmental conditions, and farming practices. Factors like heat tolerance, disease resistance, and ability to thrive on available forage and feed resources play a role in breed selection. Jerseys and Guernseys can do well in either pen-related confinement or in pasture-based situations. Brown Swiss and Ayrshires can adapt well to rugged terrains and can tolerate a range of climates.

Reproductive Efficiency

Factors such as age at first calving, fertility, and ease of calving can often come into play when selecting a breed. Breeds that exhibit low calving difficulties are often preferred.

Longevity and Health

Factors such as low susceptibility to mastitis and overall health and vigor are generally taken into account. The six most common dairy breeds generally fair well, but Ayrshires are known to be extra hardy.

Management Requirements

Different breeds may have varying management requirements. Some breeds may require specific feeding strategies, housing conditions, or specialized care. Typically, farmers assess their available resources, facilities, and management capabilities to choose a breed that aligns with their system.

Selection Criteria Comparison

Criteria	Holstein	Jersey	Guernsey	Brown Swiss	Ayrshire
Gallons / day	7–8	4–5	5–6	6–7	4–5
Butterfat content	Low	High	High	Medium	Medium
Butter & cheese	OK	Best	Best	Better	Good
Adaptability/ pasture	Good	Best	Best	Better	Better
Adaptability/ terrain	OK	Better	Better	Best	Best
Average body-weight	1500	900-1000	1000-1200	1300-1400	1000-1300
Longevity & health	Good	Better	Better	Best	Best
Milk years	3-4	4-6	4-6	3-4	3-4

Homestead Best Fit Comparison

Criteria	Best Fit
High milk quantity needs	Holstein
Moderate milk quantity needs, with limited space	Jersey and Guernsey
Moderate milk quantity needs, with large acreage available for grazing	Jersey, Guernsey, Brown Swiss, and Ayrshire
Moderate milk quantity needs, with challenging terrain available for grazing	Brown Swiss and Ayrshire

Lactation Cycle

The lactation cycle consists of the lactation period and the drying off or rest period. The lactation period is the length of time a cow is in milk production. The duration of that begins when a cow gives birth and lasts until the time the cow ceases to produce milk. The duration of the lactation period can vary among individual cows and breeds, but normally is about 305 days.

Toward the end of the lactation period, milk production may gradually decline, and the cow's energy and nutrient requirements will begin to change.

During the dry period, which typically lasts about 45 to 60 days, the cow is not milked. This dry period helps ensure the cow's health and prepares her for the next lactation cycle.

Gestation Period

The gestation period for dairy cows refers to the length of time between conception and calving. On average, the gestation period for dairy cows is about 280 to 290 days or approximately 9 months.

Some breeds may have slightly shorter or longer gestation periods. For example, Jerseys tend to have a slightly shorter gestation period compared to Holsteins. However, these differences are generally minor and fall within the range of 280 to 290 days.

It is important to provide appropriate nutrition and housing during this time to support the pregnant cow and the developing calf—especially during the dry period as the birthing date approaches, in order to facilitate the successful arrival of a healthy calf.

Breeding

If you plan to have a milk cow, you are going to need a bull. Because pregnancy is required to ultimately have a lactation cycle, the benefit of a bull is a necessary part of the equation. A viable breeding plan is therefore going to be an important part of your overall dairy endeavor. The only challenge is that it is only needed once a year.

Bull service is necessary for the milk cycle to eventually happen, but the housing and feeding of a bull year-round is a lot of expense just to benefit from the annual service. It will cost roughly $3.50 per day to keep a bull year-round. That is where a need for careful planning comes into play.

Fortunately, it is possible to have your cow serviced once a year, without the day-to-day expense of owning a bull. There are two other possibilities that can be considered—paying for bull siring service or paying for artificial insemination.

If you are fortunate enough to have a dairy somewhat close in proximity to your homestead, you might be able pay to have your cow serviced by one of the dairy bulls or have the dairy staff artificially inseminate your cow. The prices for these services can vary, but an estimate is only a phone call away. Listed below are price estimates for both services.

Cost Comparison

Criteria	Breeding Service	Artificial Insemination
Cost per cow	$40–$50	$50–$150
Success rate	90–100%	60–70%

If you don't have a dairy close by, you may be able to hire a veterinarian to do this for you. The price would be higher, of course, for the vet to arrange for the semen purchase and shipping and then for the insemination process itself.

The other option would simply be to haul your cow to a nearby bull, if one is available. This might be the cheapest, but it may not result in the ideal quality and breed you would prefer.

Breeding and Lactation Cycle Timing

Dairy cows must give birth to 1 calf per year in order to produce milk on an ongoing basis. Mature cows will come into heat about 60–90 days after giving birth. You will want to have your breeding plan ready to implement when this happens. Assuming the cow is impregnated at this time, the gestation period would be for 9 months. So, in a perfect world, the cow would be in the lactation period for about 10 months, then be in the drying off period for 2 months, and then give birth again, starting another lactation period.

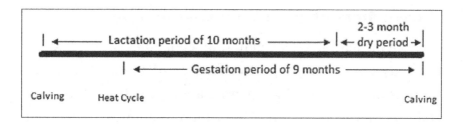

Feeding

The types of feed will generally consist of hay, pasture, and grains, and will normally be a combination of these types of feed plus minerals. The amount of feed required per dairy cow per day can vary, depending on factors such as the cow's breed, weight, stage of lactation, and nutritional requirements. These factors will dictate which combination of feed and minerals will be best at any given time.

Forage and Grains

The typical feed ration for dairy cows will normally consist of forage, such as hay, pasture grass, and/or legumes, and a mixture of grains. Grains used for dairy cow feed supplement generally incorporate a mixture of corn, oats, wheat, barley, sorghum, and/or rice.

The mixture ratio for forage and grains is typically a 60:40 ratio. These ratios can be modified to best meet the particular stage the cow is in (i.e., stage of lactation, month of gestation, etc.). Higher milk production levels generally require greater nutrient intake. As milk production increases, farmers may need to adjust the overall feed quantity to meet the cow's energy and nutritional demands.

Typical Amount of Feed per Cow per Day

A lactating dairy cow generally consumes about 2.5% of her body weight in dry matter feed per day. For example, a 1200-pound cow would consume approximately 30 pounds of dry matter feed per day. That would be about ⅓ of a 90-lb. bale of hay per day.

If you are depending on hay alone for forage, a 1200-lb. cow would require 120 bales of hay per year, or 10 bales per month. Hay and grain, even for just 1 milk cow, will require some sort of feed storage facility.

Feeding Equipment

Feeding equipment typically will consist of a feeding trough and feed bunk for containment feeding, and a trough for feeding inside the milking facility during milking.

Feed Troughs

Feed troughs can be constructed with metal, concrete, or wood. These troughs can be semicircular or square in shape. Metal troughs are often constructed by cutting a 55-gallon barrel in half, with the edges covered with angle iron.

Feed Bunks

Feed bunks are designed to hold fodder, such as hay, in the upper part, with a trough underneath to catch the hay as it is pulled out from the bunk in the feeding process. Feed bunks can be made of wooden boards or metal.

For wooden versions, the frame can be constructed with 2" x 4"s. The bunk slats extend downward from the top on both sides, toward the middle, to form a V shape. The slats are generally constructed using 1" x 4"s, about 4 ft. long, and 6–7 inches apart—wide enough for the cow to get her muzzle through. The lower trough can also be constructed with 1" x 4"s or wider lumber.

For the metal version, the frame can be constructed by welded metal pipe. The slats can consist of welded rebar, and the lower trough can be constructed by cutting 55-gallon barrels in half, and the edges covered with angle iron.

Feed Storage

Feed storage structures are used to store hay and grain. These structures can be constructed with wood or metal. Even shipping containers make excellent feed storage structures, which can be relatively inexpensive.

Watering

Daily water intake will largely depend on the stage in which the dairy cow is in. A dry cow will drink 9–12 gallons of water per day, and a lactating cow can drink up to 30–50 gallons of water per day, but the average is about 25 gallons.

Another determining factor can be the outside temperature. On average, a cow will drink 1 gallon per 100 lb. of body weight when it is cold, and 2 gallons per 100 lb. of body weight when it is hot.

Water for a lactating cow should be cool, clean, and readily available. A dairy cow will drink 50–60% of their water needs immediately after milking.

Milk Production

Milk management is the heart of your milk cow operation and will require milking, milking equipment, milk processing, and milk storage.

Milking

If you plan on having a milk cow, and milking once or twice per day, you should have a designated space for milking. This is generally inside a shed or barn, so that you are protected from the weather, and the cow can be contained while being milked.

Milking Stall

The milking space can be a stall, large enough for you and the cow, or a designated space inside a shed or barn. A trough should be available for the cow to feed while being milked, and a stanchion, which is comprised of a set of upright posts or slats—one being fixed and the other able to pivot from the bottom—in order to lock the head in place above the trough. This stanchion arrangement secures the cow and prevents her from trying to move or exit during the milking process.

Kicker Hobble Chains

One piece of equipment that is helpful during milking, especially with a young cow, is the kicker hobble chain. This chain fits around the cow's legs, just above the hock, preventing her from kicking you or knocking over the milk bucket while you are trying to milk.

Milking Stool and Bucket

Of course, the most common pieces of equipment for milking

are the stool and bucket. The stool can be made of just about any type of material, with wood the most common.

The best type of milk bucket is a stainless-steel type. This type is durable and easy to clean. It is important to keep your bucket clean and sterilized between milkings.

Milk Processing Equipment

Milk processing equipment generally includes a pasteurizer, a cream separator, a butter churn, and perhaps a cheese press.

Pasteurizer

Pasteurization is the process of sterilizing the milk to eliminate germs. This process makes the milk safer for human consumption and lengthens the time it can be safely stored. Pasteurized milk can last up to 2 weeks with proper refrigeration, while unpasteurized milk may only last 1 week.

The process of pasteurization basically consists of heating the milk and keeping it at a predetermined sterilizing temperature for a certain amount of time. The milk is then cooled for storing or further processing.

Cream Separator

As it sounds, the cream separator separates the fat from the milk, resulting in cream and low-fat milk. The cream is then used for other products, such as butter, and for cooking and baking.

Butter Churn

The butter churn is a device that agitates the cream and causes the fat globules in the cream to clump together, forming butter. A common by-product is the leftover liquid called buttermilk.

Cheese Press

The main component of milk used for cheese making is called the curd. The curd is formed by coagulating the milk, which causes it to separate into solid curds and liquid whey. The curds are then further processed to make cheese.

To begin, slowly heat the milk over low to medium heat until it reaches the desired temperature. The milk is caused to coagulate by adding an agent, such as rennet, or an acid, such as lemon juice or vinegar. This causes the milk proteins (casein) to form a curd. The liquid by-product is called whey.

The curds are then cut to allow more whey to drain. The final curd size is dependent on what kind of cheese you want to make. The curds are then heated and stirred causing more whey to separate, firming up the curds even more. Finally, the remaining whey is removed, leaving solid curds.

The solid curds are then pressed by a cheese press, forming the cheese into the particular shape you desire. The cheese is generally salted to enhance the flavor and aid in preservation. Finally, the cheese is aged to allow various enzymes and bacteria to develop and create desired flavors and textures. The longer the aging process, the firmer the cheese will be.

Whey can be used for various food items, such as ricotta cheese, mozzarella cheese, and yogurt. It can also be used in various cooking and baking recipes.

Milk Storage Equipment

The best way to store milk is in a glass or stainless-steel container in the refrigerator. For high volumes of milk, it is not uncommon to have a designated refrigerator for the sole purpose of milk storage. Of course, long-term storage should occur after the milk has been pasteurized.

Typical Maladies

Udder edema and mastitis are the most common health problems among dairy cows. Udder edema is the swelling of the udder, disrupting normal milk flow. Mastitis is an inflammation of the mammary glands.

Udder Edema

When a cow first becomes fresh after calving, the udder will become full. Ideally, the calf will be allowed to suck right away, but the newborn calf cannot consume such a large volume right after birth. It will therefore be necessary to milk the cow by hand to try to alleviate the pressure of the full udder.

The udder can be so full that it becomes hard and difficult to get the milk to flow. In such a case, gently massaging the udder with the milk or with warm water will help to soften the udder tissues and allow the udder to be emptied. This process may be necessary for several days until the udder is no longer hard, and the milk can flow without being enhanced with massaging. If edema is not addressed in a timely manner, it can lead to mastitis.

Mastitis

Mastitis causes the udder to be subject to bacterial infection beginning in the teat canal. Typical symptoms can include lack of appetite, reduced milk production, and signs of dehydration and diarrhea.

Mastitis Prevention

The best prevention of mastitis is to keep the cow, the housing facility, and the milking facility as clean and dry as possible. The use of straw from straw bales can facilitate this effort.

Mastitis Treatment

In severe cases, antibiotics might be needed to remedy the problem. In such cases, you may need to consult a veterinarian for proper treatment.

Calving and
Early Calf Management

Disease Protection

When the calf is first born, the mother's milk contains colostrum, which is high in nutrients and antibodies. A newborn calf needs these antibodies for disease protection in its digestive tract within the first 12 hours—ideally within the first 30 minutes.

If the calf does not get this needed protection within the first 24 hours, it can, and most likely will, contract a digestive tract infection called the scours. This is a serious infection and may need an antibiotic to resolve the problem. If not treated properly and in a timely manner, this infection can become fatal.

Calf/Cow Separation

Once the calf has had a chance to consume the colostrum and get well established, it will become necessary to implement a plan to separate the calf from the cow in order to save some of the milk for milking.

The cow should be placed on a schedule of being milked twice a day. For this schedule to work, the calf will need to be separated from the cow during certain hours of the day.

One common plan is to allow the calf to suck during the day, and then be separated from the cow during night, in order for the morning milk to be available for your family or vice versa.

Eventually, the calf will arrive at an age when it no longer needs milk and will need to be separated from the cow all the time. This will allow both milkings to be available for the family.

Calf Weaning

The age at which a calf can be weaned can vary. It depends mostly on your own family milk needs. The earliest a calf should be weaned is perhaps 8 weeks. If your family needs for milk are not high, then you can allow the calf to share the milk for several months.

Space Requirements

Most livestock will require some sort of containment for space management. The dairy cow will generally need a pen or corral to keep the cow close by to facilitate feeding and milking.

Pen/Corral for the Cow

The typical means for dairy cow containment is either a pen constructed with metal posts and woven wire, or with wooden posts and horizontal wooden boards.

Separate Pen for Calves

A separate pen will be needed to keep the cow and calf separated for milk management. If a pasture is accessible, the cow and calf can utilize this arrangement with one being in the pasture and the other penned, or you can allow them to graze together during the day, and separate them at night.

Happy Cow Space

The cow being a ruminant has a stomach with four compartments. During the morning feeding, dry matter is consumed and goes into the first stomach compartment called the rumen. Later in the day, after the fiber in the rumen has softened, it will be regurgitated, and the cow will rechew this regurgitated fibrous food called a cud.

The cow normally spends about 8 hours a day chewing her cuds. They normally do this while lying down and relaxing. The cow will require a flat, dry, uncrowded place where she can lay down and chew her cuds.

Shelter

Any homestead should have appropriate shelters for the various types of livestock. The milk cows should have a shed or barn available, if possible, to protect them from harsh weather and have a place to be milked. The most ideal form of dairy shelter is a wooden barn.

Other types of shelters that can be used are enclosed sheds and open-sided sheds. These structures can be made of wood or metal-framed structures with corrugated metal tops.

Pasture Management

Pasture management is a way to prevent your pasture from being overgrazed and remain vigorous and productive. Most often, the best way to accomplish this is to use a rotational grazing system—especially if you have several animals grazing, or your acreage is limited.

If a pasture is overgrazed, the grass health will diminish, and the grass will begin to die out and be replaced with less preferrable plants, such as weeds.

Grasses

There are numerous species of grass that can be introduced and used for grazing in areas with moderate-to-high rainfall. In areas where rainfall is limited, you must rely on native species.

The most common native grasses in the southwestern US are Blue Grama, Side-oats Grama, and Buffalograss. Grasses native to most other western US states are Bluebunch Wheatgrass, Idaho Fescue, and Big Bluestem. Grasses native to most eastern US states are Little Bluestem, Switchgrass, Indiangrass, Eastern Gamagrass, and Wild Rye.

Grass Growth Stages

Grasses used for grazing have three basic growth stages: vegetative phase, elongation or transitional phase, and the reproductive or flowering phase.

In the vegetative phase, the plant first uses carbohydrates stored in the root crown to form the first set of leaves. In the elongation phase, more leaves are formed. Chlorophyll develops rapidly in the young leaves, allowing photosynthesis to convert sunlight and carbon dioxide into glucose and oxygen. The glucose is used for growth and is also stored in the roots for future regrowth in the form of starch or carbohydrates.

As the grass plant continues to grow, the flowering phase of the plant develops into a seedhead. As the grass is allowed to mature, the food is stored in the roots for future new growth, and the new seed helps to maintain a healthy pasture.

Overgrazing

Overgrazing basically reverses the healthy growth trend. Continuous grazing causes the plant to continually rely on stored food for regrowth. This will eventually lead to all of the stored food in the roots to be exhausted, causing the plant to become weak and ultimately die. Nature will replace the grass plant with weeds and other undesirables.

Carrying Capacity

The best way to prevent overgrazing is to use a rotational grazing system. First, however, you must determine how many head of cattle the pasture can carry without being overused. This is called carrying capacity, which is typically represented in animal unit months (AUM). One AUM is the amount of forage that can sustain one 1000-lb. cow for 1 month.

With today's technology, two methods are most commonly used to determine carrying capacity: by estimating using charts and graphs or by using actual on-site sampling data.

Estimation Method

The estimation method includes estimating your pasture condition, and then determining forage yield by annual rainfall in inches, with the use of a forage yield chart. The result will give you an estimated yield per acre. You then multiply this by the utilization rate you want to use, which gives you the total available forage in pounds. From this figure, you can determine the total AUMs for the pasture.

Sampling Method

The sampling method is a clipping and weighing method, which is generally more applicable for small homestead applications. This will require a 0.25-square-meter frame, scissors, and paper bags, a means to dry the clipped grass, and a scale in grams.

If you have 1–2 acres, you can take 10 samples per acre. If you have multiple acres divided into pastures, then you can take 10 samples per pasture.

Begin by flagging and numbering ten locations for the samples to represent an average sampling of the area. Place the 0.25-meter frame (50 cm. x 50 cm.) on the ground at each location, and clip all the grass within the frame, close to the ground. Place the clippings in one or more paper bags and identify the bag as to location and acre, or pasture number.

After all of the locations have been clipped and bagged, arrange for the samples to be dried. After the samples have been dried, then weigh each bag in grams, and record the results. Next, add up all of the samples per acre, or per pasture, and record the total in grams. Divide the total weight by the total number of samples to get an average weight per 0.25-meter sample.

1. Multiply the average sample in grams x 35.6. = lb. per acre
2. Multiply the lb. per acre x number of acres = total lb.
3. Multiply the total lb. x desired utilization rate (i.e., 50%) = adjusted lb.

4. Adjusted lb. ÷ 26 lb. per day = animal unit days (AUD)

5. AUD ÷ 30 days = animal unit months (AUM)

The AUM figure can then be used to determine how to arrange your rotational grazing system. For instance, let's say you have 4 acres and a total of 12 AUMs. That means you can have 12 cows on 4 acres for 1 month, or you can have 1 cow on 4 acres for 12 months, or any combination equaling 12 AUMs.

Rotational Grazing

Without rotational grazing, the pastures are generally grazed unevenly, meaning some areas are grazed more heavily than others. The main benefit of rotational grazing is to allow the pasture to be grazed more uniformly, and to allow the grass time to regrow and have a rest period before being grazed again. This allows the grass to recover and replace the needed energy in its root system.

Using the example above of 4 acres having a carrying capacity of 12 AUMs, the acreage could be divided into pastures. If the choice were to have 1 cow on 4 acres for 12 months, the acreage could be split into 2, 3, or 4 pastures (i.e., 2 pastures for 6 months each; 3 pastures for 4 months each; or 4 pastures of 3 months each).

Obviously, the more pastures created, the more fencing is required. This can also be achieved with electric fencing, reducing time, labor, and materials. In a general sense, the more pastures you have, the less impact on the soil from grazing you will have, but water availability may govern that decision.

Please note that 1 AUM = one 1000-lb. cow. This can be converted to sheep by dividing the AUMs by the animal unit equivalent (AUE). The AUE for sheep is .2.

Expenses

Purchase Cost

The purchase cost for a milk cow in the US can vary quite a bit—from $1000 to $5000, with the average being about $2000–$2500. The price largely depends on the breed, age, weight, and confirmed lineage. Dairy cows are generally sold by weight, with prices averaging $150–$200 per 100 wt. (cwt.).

Breeding Costs

As discussed earlier in this chapter, a bull service will be required annually in order to keep the milk cycle going. This can be done by owning a bull, but it is much cheaper to pay an annual fee for breeding service. This cost is $40–$50 for a siring service and $50–$150 for artificial insemination (AI).

Feed Cost

Feed for a milk cow will generally consist of both hay and grain. These prices can vary widely depending on the area and time of year.

Hay Cost per Bale

Depending on the area, alfalfa hay costs $200–$245 per ton. That would translate to $9–$11 per 90-lb. bale, with an average of about $10 per bale.

Grain Cost per 50-Lb. Bag

Livestock grain usually comes in a 50-lb. bag, and can contain a mixture of ground corn, sorghum, wheat, barley, oats, etc.—with molasses sometimes included. The average cost for a 50-lb. bag generally runs about $12–$15, with an average cost of $13.75.

Daily Feed Cost

An average 1200-lb. cow will eat about 30 lb. of hay per day, which would be ⅓ of a bale per day, or $3.30 per day.

In addition, a lactating cow should also have grain as part of the diet. One way to figure how much grain mix to feed per day is by using the following formula: .6 lb. grain mix per 3 lb. of daily milk.

Using a daily milk volume of 4 gallons, you can determine the daily usage and cost.

- 4 gallons per day x 8.6 lb. per gallon = 34.4 lb. of milk
- 34.4 ÷ 3 = 11.5 lb.
- 11.5 lb. x .6 lb. grain mix = 7 lb. of grain mix per day, or 3.5 lb. per feeding
- 50-lb. grain bag ÷ 7 lb. per day = 7 days
- $13.75 ÷ 7 days = $2 per day for grain

Hay cost =	$3.30 per day
Grain cost =	$2.00 per day
Total cost =	$5.30 per day for feed

Feeding Equipment

Feeding equipment such as feed troughs and feed bunks can be purchased or fabricated on-site. Such items can be fabricated out of a variety of materials, but for long-term usage, troughs and feed bunks are typically fabricated using metal, wood, or concrete.

Troughs

On average, a feed trough will cost roughly $20 per linear foot. The typical size is 18 in. wide, 12 in.–14 in. deep, and 8 ft.–10 ft. long.

Feed Bunks

Feed bunks will run $60–$80 per linear foot. The typical size is about 4 ft. wide, 5 ft. long, and 5–6 ft. high.

Watering Equipment

Water containers can come in a variety of shapes, sizes, and materials. The most common material is either plastic or galvanized metal. The common shapes are either round or oblong.

Plastic Water Troughs

A 100-gallon plastic oblong-shaped watering trough will cost $100–$150 ($1–$1.5 per gallon). Larger troughs will run closer to $2 per gallon.

Galvanized Water Troughs and Tanks

An oblong water trough, with the dimensions of 2 ft. x 2 ft. x 6 ft. (169 gallons), will run about $150 ($.89 per gallon). A round galvanized water tank of 2 ft. x 6 ft. (389 gallons) costs $450 ($1.16 per gallon).

Milking Facility

You should have a shed or part of a barn accessible to function as a milking stall. The stall should include a small feed trough and a stanchion. Simple sheds can cost $15 per sq. ft.

Metal barns can cost $25–$50 per square foot. Wooden barns, on average, cost $45–$65 per square foot to build.

Milk Processing Equipment

Milk processing equipment can include a pasteurizer, a cream separator, a butter churn, and a cheese press. These milk processing steps can be done by hand, by using common household equipment and/or using homemade equipment.

However, if you plan on processing medium-to-large volumes of milk on a regular basis, it may be wise to purchase the equipment—especially items like the pasteurizer that require high temperature control for killing germs.

The pasteurizer will cost about $750; the cream separator $150–$350; a butter churn $350–$450 (a small manual 1-gallon churn can be purchased for as little as $75 but will require a moderate amount of time and work). A cheese press will cost $150–$250.

Milk Storage

Generally, a moderate-to-high volume of daily milk production will require a designated refrigerator for the sole purpose of milk storage. An average size refrigerator costs $350–$500.

Containment Costs

The cost of fencing is another mandatory expense. On average, corrals or pens around the heart of the homestead where the animals are kept and fed are generally made of wood or welded pipe, and pasture fencing is normally constructed with metal posts and woven wire.

Wooden Fence Cost per Foot

Split rail wooden fences will cost $8–$12 per linear foot. Labor will cost ~$10 per linear foot.

Pipe Fencing Cost per Foot

Pipe fencing will cost $10–$15 per linear foot. This is more costly than a wooden corral but will last much longer as well.

Woven Wire Fence Cost per Foot

The average cost for a woven wire fence with metal posts will depend on the type of mesh but on average will run about $1.50–$1.90 per linear foot. This type of fencing is much more cost-effective for pasture fencing.

Electric Fencing

Electric fencing for livestock generally runs about $3–$4 per linear foot. This type of fencing can be a good choice for pasture divisions when livestock are moved on a monthly or bimonthly basis.

Bedding Cost

Bedding should be provided in the shed or barn during the winter and calving season. This can be facilitated with straw. One straw bale will cost approximately $5 per bale, and one bale should last a week.

Vet Care

It is not uncommon to require a veterinarian's service occasionally. You might need treatment for such issues as mastitis, calf scours, or to assist with AI. It, therefore, may be wise to include this as a possible expense item in your annual budget.

Start-Up Expense Summary Charts

The following expense summaries are designed to provide a simple example for demonstrating start-up expenses. The following charts will be based on initial purchase of a 1200-lb. milk cow, for a homestead with 1 acre of pasture. The corral will be 60 ft. x 40 ft., with two 20 ft. x 20 ft. interior pens. An enclosed shed (1000 sq. ft.) will be used for shelter and milking. The 1 acre of pasture will be enclosed with woven wire fencing.

Initial Livestock Investment Expense Chart

Item	Cost
1200-lb. cow	$2400

Infrastructure and Equipment Expense Chart

Item	Cost
Wooden corral 60 ft. x 40 ft.	200 ft. @ $10 / ft. = $2000
Two interior pens 20 ft. x 20 ft.	60 ft. @ $10 / ft. = $600
One-acre perimeter fenced with woven wire fence & metal posts	836 @ $1.75/ft. = $1463
Milking shed (1000 sq. ft.)	@ $15/sq. ft. = $15,000
Shipping container for feed	$2000
Feed trough (8 ft.)	$160
Feed bunk (5 ft.)	$400
Water trough (100 gallon)	$100
Total	**$21,723**

Milk Processing Equipment

Item	Cost
Pasteurizer	$750
Cream separator	$150
Butter churn	$350
Cheese press	$150
Refrigerator	$350
Total	**$1750**

Operating Expenses Summary Chart

The operating expenses are based on milking and maintenance for a 1200-lb. milk cow, for a homestead with 1 acre of pasture.

Feed Costs

Item	Daily Cost	Monthly Cost
Alfalfa	$3.30	$99
Grain	$2.00	$60
Minerals	$.10	$3
Total	**$5.40**	**$162**

Animal Care

Item	Annual Cost
Bedding	$260
Breeding cost	$100
Vet care	$150
Total	**$510**

Benefits and Revenue

The value of having a milk cow can be threefold: (1) an abundant supply of milk and milk products for your family, (2) a leftover supply of milk and milk products to sell, and (3) a side benefit of teaching children how to be responsible and develop a good work ethic.

Milk Value

Many of the health benefits of having fresh milk for your family can be immeasurable. One of the more quantifiable benefits is perhaps the monetary savings, but this is largely governed by the amount of milk and milk products your family consumes.

You should weigh all of the costs and benefits to accurately discern if having the expense and responsibility that comes with owning a milk cow is a good fit for you and your family.

The price of milk in most areas of the US is $4.30 per gallon (2024 pricing). If your cow produces 4 gallons per day, that is a $17.20 value in today's prices. If you turn some of that milk into milk products, this value can increase.

Milk Products

Milk products include mostly butter and cheese. The by-product of making butter is buttermilk. The by-product of making cheese is whey. All of these products and by-products can be used by your family or sold if you have a local demand.

Butter

One gallon of milk will usually yield 1–1.5 pints of cream. The cream can be churned to make approximately ⅓ to ½ lb. of butter.

Cheese

Ten pounds of milk (1.2 gallons) can yield approximately 1 pound of hard cheese or 1.25 pounds of soft cheese (i.e., mozzarella, ricotta, cottage cheese, etc.).

By-Products

The by-products from milk products, such as buttermilk and whey, can be used or sold locally.

Buttermilk

In the process of making butter, about 25% of the weight of milk will end up as buttermilk. If 1 gallon of milk will produce ½ pound of butter, the by-product will be about 2 pounds, or 1 quart, of buttermilk.

Whey

In the process of making cheese, 10 pounds or 1.2 gallons of milk will produce 1 pound of cheese. Therefore, about 9 pounds, or a little over 1 gallon of whey will result as the by-product.

Calf Revenue

If your cow is able to produce 1 calf per year, you can sell the calf for a profit. You should try to keep the calf on mother's milk for at least 8 weeks, but after weaning, you could sell the calf at an age of 3–4 months.

A calf that is 3–4 months old should weigh approximately 350 lb. At 2024 prices of $155 per cwt., a 350-lb. calf can bring about $542.50 ($155 x 3.5).

Cost/Benefit

The purpose for this analysis is to provide a way to examine the cost versus the benefit of owning a milk cow. By now, you have been able to see that owning and operating a milk cow operation on your homestead is relatively expensive, with a lot of responsibility.

In a general sense, most homesteads choose to go the milk cow route when they have one or both of the following: (1) a big family that can help with the work, and utilize large quantities of milk, or (2) have a neighboring community with whom you can share the expense, or to whom you can sell the unused milk and milk products.

Savings/Loss

The following chart compares the feed and care cost to the milk yield value. The benefit-to-cost ratio will use annual figures in the comparison.

Please note that cow purchase cost and infrastructure cost are not included. The same infrastructure can be used for other livestock options.

The savings/loss figure reflects more of a savings from having to purchase your milk and milk products (which can typically be twice as much), but isn't taking into account the labor involved in milking the cow and processing the milk.

Item	Annual Expense	Annual Value or Revenue	Annual Savings/Loss
Feed & misc.	$1944		
Animal care	$510		
Total expense	**$2454**		
Milk value		$5160	
Calf revenue		$542	
Total savings/ revenue		**$5702**	
Savings/loss			**+ $3248**

• Based on the figures above, the benefit-to-cost ratio is 2.3:1.

Picture Gallery

Holstein

Jersey

Brown Swiss

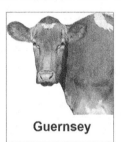

Guernsey

Ayrshire

CHAPTER 2

Milk Goats

Most homesteaders who want to produce their own milk generally use milk goats in lieu of milk cows because goats, being smaller, require less space and eat less feed. It is actually a great fit for those who don't require a lot of milk or have limited acreage.

This chapter will compare the most common dairy goat breeds and list the various differences and similarities regarding size, weight, temperament, and milk production. Pros and cons will also be listed to help discern the best fit for your homestead.

Basic requirements regarding needed space and shelter will be analyzed, as well as daily feed costs, daily milking options, kid management, and milk management. This chapter will conclude with a cost/benefit analysis in order to assist you in the decision-making process.

Dairy Breeds

There are many varieties of milk goats to choose from in the US for homesteads. The most common five breeds are listed below.

Nigerian Dwarf

The Nigerian Dwarf breed is a popular choice among small farms and homesteads due to their small size and gentle temperament. Their milk production is low to moderate, but the butterfat content is the highest of the breeds. Their small size allows them to be a good fit for limited acreages and small families. Originating from West Africa allows this breed to tolerate higher temperatures. Their color is usually black or brown.

Pros	Cons
• Small acreage needs	• Low milk production
• Low feed requirement	
• High butterfat content	

Alpine

The Alpine milk goat is another favorite among homesteaders because of its gentle and friendly nature. It is a high milk producer with a medium-to-high butterfat content. Originating from the French Alps, the Alpine breed is known for its versatility and adaptability and can adapt well to colder temperatures. The Alpine can come in a range of solid colors.

Pros	Cons
• Sweet nature	• High energy
• Adaptable	• Higher feed needs

Saanen

The Saanen is one of the larger breeds and is a high milk producer making this breed a favorite for homesteads with higher milk volume needs. Being a larger breed, the Saanen will require more space and higher fences. Even though this breed produces a lot of milk, the butterfat content is at the low end. The Saanen is white in color and originates from Switzerland.

Pros	Cons
• Calm and quiet nature	• Low butterfat content
• Highest milk production	• Milk less creamy

Nubian

The Nubian is another large breed and can fill a dual purpose of both milk and meat. Milk production is moderate with about 1 gallon per day, but the butterfat content is one of the highest among dairy goat breeds. This breed comes with a mixture of colors, has large floppy ears, and has a curious nature. On the downside, the Nubian has a reputation of being loud. Originating from Africa, this breed can thrive in more hot and arid climates.

Pros	Cons
• High butterfat content	• Loud
• High heat tolerance	• Higher feed needs
• Dual purpose	

LaMancha

This breed is a common choice for homesteads looking for high milk production. The milk also has a high butterfat content and a mild flavor. The LaMancha is a medium-to-large goat with uniquely small ears. It has an easy-going personality and is generally healthy and hardy. The color is generally brown.

Pros	Cons
• Easy to manage	• Strange ear appearance
• High milk production	
• Second highest butterfat	

Other Breeds

The Toggenburg and the Oberhasli are two additional dairy goat breeds that are used for milk in the US. Both originate from Switzerland. They both are medium-sized and hardy and are good milk producers, with medium-to-high butterfat content.

Breed Selection Considerations

When choosing a milk goat for your homestead, several factors are generally considered to ensure that your specific needs will be best served. Here are some of the most common considerations.

Milk Production

One of the most common things to consider when choosing a milk goat is the volume of milk produced. If high milk production is a priority, the Saanen, Alpine, and LaMancha breeds are good choices. If low volumes are a preference, the Nigerian Dwarf would be a good fit.

Butterfat Content

Butterfat content is another high priority consideration for those interested in making cheese, butter, and other milk products. The milk goats with the highest butterfat content are the Nubian and the Nigerian Dwarf.

Temperament

Almost all of the milk goat breeds are easy-going and have a mild temperament. The common favorites for their sweet and gentle personalities are the Saanen, the Alpine, and the LaMancha.

Number Requirements

Because goats are herd-oriented animals, you should have at least 2 as a bare minimum, but 3 as a minimum is often recommended. Obviously, the maximum numbers will be contingent on space availability.

Space Requirements

Space requirements are an important criterion to consider. Goats will require moderate space for shelter and larger space for grazing and/or typical daily activity (exercise).

On average, each dairy goat requires about 15–20 sq. ft. for shelter. This needs to be a three-sided structure. In addition, dairy goats, like all goats, are active animals and require adequate space to stay healthy and content. On average, a dairy goat needs at least 200 square feet of open space per goat for outside daily activity.

If goats have access to pasture, the more space you can provide the better. Pasture not only allows for exercise but also provides a natural and varied diet, which is beneficial for their health and milk production.

However, if pasture is not available, you can still create a stimulating environment for the goats by providing climbing structures and platforms to play on.

It's important to note that these space requirements can vary based on factors such as the breed of the goats and their age. Providing ample space for the goats will help prevent stress, behavioral issues, and health problems, resulting in happier and healthier dairy goats and better-tasting milk.

Feed Requirements

Feed requirements are essential for a healthy animal and optimum milk production. These requirements will vary according to breed, age, and stage of lactation.

Generally, milk goats require a mixture of hay and grain, but they are also browsers, meaning they like to eat browse (woody material like tree branches) and forbs (weeds).

Dual-Purpose Possibility

Some breeds can serve a dual purpose, providing both milk and meat opportunities by way of their offspring. The larger breeds are better suited for this dual purpose. The Nubians are a good candidate.

Selection Criteria Comparison

Criteria	Nigerian Dwarf	Alpine	Saanen	Nubian	LaMancha
Ht. (in.) doe/buck	22/23	30/32	30/32	30/35	28/30
Wt. (lb.) doe/buck	75/80	135/170	135/160	135/175	130/160
Temperament	Gentle	Friendly	Sweet	Friendly	Sweet
Lactation days	305	288	305	288	290
Gestation days	145–150	197	148–156	145–155	145–155
Gallons/day	.5	1–2	1.5–3	1	1
Butterfat content	6–10%	3.5%	3%	5%	4.2%
Butter & cheese	Best	Good	Good	Better	Better
Dual purpose	OK	Better	Better	Best	Good

Homestead Best Fit Comparison

Criteria	Best Fit
High milk quantity needs	Saanen
Moderate milk quantity needs, with limited space	Nigerian Dwarf
Moderate milk quantity needs, with large acreage available for grazing	Alpine, Nubian, LaMancha
Moderate milk quantity needs, with challenging terrain available for grazing	Alpine, Toggenburg, Oberhasli

Lactation Period

The lactation period is the length of time a milk goat is in milk production. The duration of the lactation period begins when the goat gives birth—an event called kidding or freshening.

The lactation period lasts until the time the goat "dries up." Sometimes, this drying up needs to be intentionally induced in order to give the doe a rest period of about 2 months. The duration of the lactation period can vary among individual does and breeds, but normally is 288–305 days.

Gestation Period

The gestation period for dairy goats refers to the length of time between conception and giving birth. On average, the gestation period for milk goats is about 145 to 197 days or approximately 5–6.5 months.

Breeding

The service of a male goat (buck) will be needed in order to maintain a consistent, ongoing milk cycle. A choice needs to be made to either have a buck as part of your milk operation or make arrangements to have your does serviced by a borrowed buck.

The decision largely depends on your herd size. If you have 3 does or more, it may be more feasible to own a buck. The purchase price for a buck will range from $150 to $500, and on average will cost about $5 per week or $260 per year for feed.

If you have only 1–2 does, it may be more cost-effective to pay for the breeding service. The breeding service to use a borrowed buck will cost about $35–$50 per doe. AI service will run about $50–$150 for semen, storage, and vet costs.

There are several things to be aware of, if you choose to own a buck. The time of breeding needs to be managed, requiring the buck to be kept separated from the does all the time, except for the brief period of servicing. This will require a year-round separate pen for the buck. Another drawback is that the buck is smelly.

Breeding and Lactation Cycle Timing

A doe will come into heat for a duration of 12–36 hours, every 18 to 21 days. If you choose to use a breeding service or AI, you will need to plan ahead and be ready when the doe comes into heat.

Because the gestation period is 5–6.5 months, it is possible to arrange for your doe to "freshen" twice a year. However, most homesteads have more than 1 doe and try to stagger the kidding times, so that milk is available year-round with only one annual freshening per doe. It is common to have some does bred in the fall in order to deliver in the spring, and others bred in the spring to deliver late summer/early fall. This schedule allows the majority of the extra milk to be produced in the summer, making it easier to share the milk with their kids when there is plenty of forage in the pasture.

Feeding

The types of feed will generally consist of hay and grain and will normally be a combination of these types of feed plus minerals. The amount of feed required per milk goat per day can vary depending on factors such as the goat's breed, size, stage of lactation, etc.

Forage and Grains

The typical feed ration for milk goats will normally consist of hay and grain. Grass or clover hay can be fed in bale form or alfalfa in a pellet form. Grains fed as supplement generally incorporate a mixture of corn, oats, barley, peas, soybeans, and cottonseed.

Typical Amount of Feed per Goat per Day

On average a lactating doe will require good quality hay and grain with a protein content of about 12–15% for hay and 16–18% for grain. Total daily intake of dry matter should be based on roughly 3–4% of their body weight. For instance, a body weight of 150 lb. would require approximately 5 lb. of forage per day.

The amount of grain should be formulated by using 1 lb. of grain per 3 lb. of daily milk produced. If a goat is producing 1 gallon per day, that would be 8.6 lb. of milk ÷ 3 = 2.8 lb. of grain per day.

Total intake, therefore, would be roughly 2.5-lb. dry matter, and 1.4-lb. grain per feeding. Because goats are browsers by nature, the dry matter should be fed in bale form and should be provided in a feed bunk to reduce waste. Grains should be fed in a trough.

Feeding Equipment

Feeding equipment will normally consist of a feeding trough and feed bunk, for containment feeding, and a trough or bucket, for feeding inside the milking facility during milking.

Feed Troughs

Feed troughs are generally made of metal, wood, or plastic. These troughs can be semicircular or square in shape. They can be handmade or purchased.

Feed Bunks

Goats do not like eating hay off the ground. Therefore, feed bunks are highly recommended for feeding hay in order to prevent waste. Even with the use of a feed bunk, some hay will end up being dropped on the ground and trampled. To avoid this waste, the feed bunk should be sized specifically for your goats.

Feed bunks are designed to hold fodder such as hay in the upper part, with a trough underneath to catch the hay as it is pulled out from the bunk in the feeding process. Feed bunks can be made of wooden boards or metal.

For wooden versions, the frame can be constructed with 2" x 4"s. The bunk slats that extend downward from the top on both sides, toward the middle to form a V shape, are generally constructed using 1" x 4"s, about 4 ft. long and 3–4 inches apart—wide enough for the goat to get their nose through. The lower trough can also be constructed with 1" x 4"s or wider lumber.

For the metal version, the frame can be constructed by welded metal pipe. The slats can consist of welded rebar, and the lower trough can be constructed by cutting 16-gallon barrels in half, and the edges covered with angle iron.

Both wooden and metal feed bunks are becoming more available for purchase at your local feed store or online.

Feed Storage

Feed storage structures are used to store hay and grain. These structures can be constructed with wood or metal. Even shipping containers make excellent feed storage structures, which can be relatively inexpensive.

Watering

On average, a lactating goat will drink 3.5 gallons of water for each gallon of milk produced. Non-lactating goats will drink 2–3 gallons per day.

Milk Production

Milk management is the heart of your milk goat operation and will require milking, milking equipment, milk processing, and milk storage.

Milking

The process of milking a goat is similar to milking a cow by hand. The difference is that the milk goat only has two teats, and they are generally larger in size than a cow. Like the milk cow, the milk goat will need to be milked twice a day.

You will need to begin by cleaning the udder and teats with warm water and a rag. You can store both in a bucket under the milk stand or table.

Once the udder and teats are clean, you can begin milking. You will need a stainless-steel bucket to hold the milk. You begin by placing your fingers around the teat, as close to the udder as possible. Then first squeeze the forefinger and thumb tightly and hold that, as you then squeeze the remaining fingers in sequence, from top to bottom. Then repeat the process with each hand, over and over, until the udder is empty.

Once finished, it is then recommended to dip each teat in a teat-sterilizing solution. You can use hydrogen peroxide or bleach at 1 oz. per quart of water.

Milking Equipment

Goat milking will require a milking stand or table. This table is generally kept in a milking stall for livestock separation purposes. The table can be handmade or purchased.

The table can be constructed with wood or metal. The dimensions for the tabletop are roughly 2 ft. x 4 ft. The legs would be 20"–24" tall, so that you can sit while milking. A stanchion will need to be constructed on one end of the tabletop to hold the goat in place while milking.

The stanchion can consist of two upright wooden 2" x 4"s— one that is fixed in place, and one that pivots at the bottom, so that it can slide over at the top and be fastened to the one that is fixed, in order to lock the head in place, preventing movement. Just below the stanchion would be a container for holding grain for the goat to eat while being milked.

Milk Processing

Milk processing equipment generally includes a pasteurizer, a cream separator, a butter churn, and perhaps a cheese press.

Pasteurizer

Pasteurization is the process of sterilizing the milk to eliminate germs. This process makes the milk safer for human consumption and lengthens the time it can be safely stored. Pasteurized milk can last up to 2 weeks with proper refrigeration, while unpasteurized milk may only last 1 week.

The process of pasteurization basically consists of heating the milk and keeping it at a predetermined sterilizing temperature for a certain amount of time. The milk is then cooled for storing or further processing.

Cream Separator

As it sounds, the cream separator separates the fat from the milk, resulting in cream and low-fat milk. The cream is then used for other products, such as butter, and for cooking and baking.

Butter Churn

The butter churn is a device that agitates the cream and causes the fat globules in the cream to clump together forming butter. A common by-product is the leftover liquid called buttermilk.

Goat butter is a rich and flavorful product with a slightly different taste compared to cow's milk butter. Some people prefer goat butter for its unique taste and properties.

Cheese Press

The main component of milk used for cheese making is called the curd. The curd is formed by coagulating the milk, which causes it to separate into solid curds and liquid whey. The curds are then further processed to make cheese.

Pour the fresh goat's milk into a clean stainless-steel pot. Slowly heat the milk over low-to-medium heat until it reaches the desired temperature. The specific temperature will depend on the type of cheese you want to make.

Add an agent, such as rennet, or an acid, such as lemon juice or vinegar, then let the milk sit undisturbed for some time (usually about an hour) to allow it to coagulate and thicken. If using lemon juice or vinegar, the milk should start to curdle immediately.

Once the milk has coagulated, cut the curds into small pieces using a long knife. The size of the curds will depend on the type of cheese you're making. Slowly heat the curds while stirring gently. The heat helps to firm the curds as they drain off more whey.

Once the curds have firmed, you can use a colander lined with cheesecloth or a cheese mold with draining holes to separate the curds from the whey. After draining the whey, you can sprinkle some cheese salt over the curds and gently mix it in. The amount of salt will depend on your taste preference.

If using a cheese mold, press the curds into the mold to give the cheese its shape. Alternatively, if making a soft cheese, you can simply transfer the curds to a container for aging. If you're making a cheese that requires aging, place the cheese in a cool, humid environment, such as a cheese-aging container or a refrigerator. The aging time will depend on the type of cheese and your desired taste.

Milk Storage

It is important to chill the milk as soon as possible after milking to prevent the milk from taking on a strong goat flavor. The best way to store milk is in a glass or stainless-steel container in the refrigerator. For high volumes of milk, it is not uncommon to have a designated refrigerator for the sole purpose of milk storage. Of course, long-term storage should occur after the milk has been pasteurized.

Kidding and Kid Management

It is not uncommon for goats to have twins—sometimes triplets. It will be important to allow the kids to have access to the doe's milk as soon as they are able to nurse. For the first couple of weeks, the milk will contain colostrum. This colostrum contains nutrients and antibodies, which will help the goat kid to develop its immune system to fight diseases.

Kid/Doe Separation

After the kids have been allowed to nurse for the first couple of weeks, it will be important to arrange a schedule to share the milk.

The doe should be placed on a schedule of being milked twice a day. For this schedule to work, the kids will need to be separated from the mother during certain hours of the day.

One common plan is to allow the kids to nurse during the day, and then be separated from the doe during the night, in order for the morning milk to be available for your family—or vice versa.

Eventually, the kids will get to an age when they no longer need milk and can be separated from the doe full-time, allowing both milkings to be available for the family.

Kid Weaning

The age at which kids can be weaned can vary. It depends mostly on your own family milk needs. The typical age for kids to be weaned is 6–8 weeks. Another way to determine the proper time for weaning is to use weight. The proper weight for weaning would be roughly 2—2.5 x the birth weight.

Containment

Milk goats will require some form of containment when being used for milking. Generally, the milk goats and their kids are kept in one large pen during the day and then separated at night.

Communal Feeding Pen

The large communal feeding pen for does and kids can be constructed of a variety of materials. Generally, woven wire fencing material is used. If wood or pipe is used to construct the large pen, woven wire will still need to be added to contain the goats.

Goats will test the fence by leaning or trying to climb under, over, or through any opening. With that in mind, the fence should include woven wire (not welded), with the square openings no larger than 4" x 4".

Kidding Stalls

Stalls for the does giving birth (kidding) are typically constructed in some sort of shelter, like an enclosed shed or barn. The size of the stalls can vary, but the typical size ranges from 4 ft. x 6 ft. to 5 ft. x 7 ft. The back and sides of the stall are normally constructed with ½" to ¾" plywood, 4 ft. tall, with a front gate made with woven wire fencing material, framed with wooden 2" x 4"s.

Pasture Fencing

Fencing for pastures is generally constructed with woven wire fencing and metal posts. The square size of the woven wire mesh should be no larger than 4 inches. The height of the fence should be 4–5 ft. tall.

Internal pasture divisions can be constructed with woven wire fencing or by using electric fencing.

Shelter

Any homestead should have appropriate shelters for the various types of livestock. The goats should have a shed or barn available, if possible, to protect them from harsh weather, have a place to be milked, and have a place for bedding down at night.

Goats are a communal, herd-type animal that prefer to bed down together. It's not uncommon to have a shed or small barn with shelves built into the walls, at different levels, for the goats to hop up onto and sleep on the pallet-sized shelves.

The most ideal form of dairy goat shelter is a wooden barn. Other types of shelters that can be used are enclosed sheds. These structures can be made of wood or metal-framed structures with metal siding and corrugated metal tops.

Health Precautions/Maintenance

On average, milk goat breeds are relatively healthy and easy to maintain, but like any dairy animal, it is necessary to take necessary precautions as preventative maintenance.

Some of the possible lactation-related ailments include udder edema, mastitis, and ketosis. Other possible ailments can include listeriosis, enterotoxemia, coccidiosis, and worms.

The lactation-related ailments can be addressed with the appropriate treatment when symptoms arise. The non-lactation-related ailments can generally be avoided by preventative maintenance injections or feed supplements.

Udder Edema

Udder edema is a swollen udder-related problem causing the udder to be caked and hard shortly after freshening, preventing milk flow. This problem can be resolved by massaging the udder with warm water, until the milk can be extracted by hand milking.

Mastitis

Mastitis is a bacterial problem in the teat canal, generally experienced during early weeks of lactation. The problem can be resolved with antibiotics. A veterinarian may need to be consulted for proper treatment.

Lactational Ketosis

Lactational ketosis is a metabolic disorder, brought about due to a nutritional imbalance. This can be an issue of having too much hay and not enough grain during late pregnancy and early lactation. Possible remedies can include gradually reducing the forage intake and gradually increasing the grain intake.

Listeriosis

Listeriosis is a bacterial problem that can be caused from eating moldy hay. Typical remedies include an antibiotic treatment. A veterinarian may need to be consulted for proper treatment.

Enterotoxemia

Enterotoxemia is commonly called "overeating disease." Anti-inflammatories, antitoxins, activated charcoal, and probiotics can be helpful during the early onset of the disease. A veterinarian may need to be consulted for proper treatment.

Coccidiosis

Coccidiosis is an intestinal problem caused by tiny intestinal parasites. The common symptom is diarrhea and can usually be remedied by administering Banamine. A veterinarian may need to be consulted for proper treatment.

Worms

Common worm problems can be treated as preventative maintenance by using a deworming product found at your local feed store.

Expenses

Purchase Cost

The purchase cost for a milk goat in the US can vary quite a bit—from $100 to $800, with the average being about $150–$500. The price largely depends on the breed, age, weight, and documented lineage.

Breeding Costs

As discussed earlier in this chapter, a buck service will be required annually in order to keep the milk cycle going. This can be done by owning a buck, but can be much cheaper to pay an annual fee for breeding service if you have fewer than 3 does. This breeding service can cost $35–$50 for a borrowed buck or $50–$150 for artificial insemination (AI).

Feed Cost

Feed for a milk goat will generally consist of both hay and grain. These prices can vary widely depending on the area and time of year.

Hay Cost per Bale

It is OK to feed goats grass hay in lieu of the more expensive alfalfa hay. Grass hay costs, on average, 30% less than alfalfa. Depending on the area, alfalfa hay costs $200–$245 per ton, and grass hay is about $140–$240 per ton. That would translate to $9–$11 per 90-lb. bale for alfalfa, with an average of about $10 per bale. Grass hay is roughly $7 per bale. Again, these prices can vary widely per geographic area and time of year.

Grain Cost per 50-Lb. Bag

Livestock grain usually comes in a 50-lb. bag and can contain a mixture of ground corn, sorghum, wheat, barley, oats, etc.— with molasses sometimes included. The average cost for a 50-lb. bag generally runs about $12–$15, with an average cost of $13.75.

Daily Feed Cost

An average 150-lb. doe will eat about 5 lb. of forage per day, which would be a tenth of a bale of grass hay per day, or $.70 per day. This is based on a 50-lb. bale of grass hay at $7 per bale. The weight per bale and price per bale can vary widely depending on the area, time of year, and other economic factors.

In addition, a lactating goat should also have grain as part of the diet. The amount of grain should be formulated by using 1 lb. of grain per 3 lb. of daily milk produced. If a goat is producing 1 gallon per day, that would be 8.6 lb. of milk ÷ 3 = 2.8 lb. of grain per day.

- 50-lb. grain bag @ $13.75 = $.28 per lb.
- $.28 x 2.8 lb. per day = $.78 per day

Hay cost = $.70 per day

Grain cost = $.78 per day

Total cost = $1.48 per day for feed ($.74 for Nigerian
              Dwarf)

Feeding Equipment

Feeding equipment such as feed troughs and feed bunks can be purchased or fabricated on-site. Such items can be made out of a variety of materials, but for long-term usage, troughs and feed bunks are typically fabricated using metal or wood. Metal or plastic versions can be purchased online.

Troughs

On average, a feed trough will cost roughly $20–$40 per linear foot. The typical size is 12 in. wide, 8 in.–9 in. deep, and 6 ft. long.

Feed Bunks

Feed bunks will run $60–$100 per linear foot. The typical size is about 4 ft. wide, 5 ft. long, and 5 ft. high.

Feed Storage

Feed storage can be in a shed, barn, or shipping container. A good 8 ft. x 40 ft. shipping container can be purchased for $2000.

Bedding Cost

Bedding should be provided in the shed or barn during the winter and kidding season. Each kidding pen especially, should be a clean and dry space with bedding. This can be done with straw bales. One straw bale cost approximately $5 per bale, and one bale should last a week.

Watering Equipment

Water containers can come in a variety of shapes, sizes, and materials. The most common material is either plastic or galvanized metal. The common shapes are either round or oblong.

Plastic Water Troughs

A 40-gallon plastic oblong-shaped watering trough will cost $60 ($1–$1.5 per gallon). Larger troughs will run closer to $2 per gallon.

Galvanized Water Troughs and Tanks

An oblong water trough, with the dimensions of 2 ft. x 2 ft. x 4 ft. (44 gallons) will run about $110 ($2.5 per gallon). A round galvanized water tank of 2 ft. x 3 ft. (53 gallons) costs $120 ($2.27 per gallon).

Milking Facility and Shelter

You should have a shed or part of a barn accessible to include a milking stall. The stall should include a milking stand or table, with a small feed trough and a stanchion. Simple sheds can cost $15 per sq. ft. Metal barns can cost $25–$50 per square foot. Wooden barns, on average, cost $45–$65 per square foot to build.

Milk Processing Equipment

Milking processing equipment can include a pasteurizer, a cream separator, a butter churn, and a cheese press. These milk-processing steps can be done by hand, by using common household equipment and/or using homemade equipment.

However, if you plan on processing medium-to-large volumes of milk on a regular basis, it may be wise to purchase the equipment—especially items like the pasteurizer that require high temperature control for killing germs.

The pasteurizer will cost about $750; the cream separator $150–$350; a butter churn $350–$450 (a small manual 1-gallon churn can be purchased for as little as $75 but will require a moderate amount of time and work). A cheese press will cost $150–$250.

Milk Storage

Generally, a moderate-to-high volume of daily milk production will require a designated refrigerator for the sole purpose of milk storage. An average size refrigerator costs $350–$500.

Containment Costs

The cost of fencing is another mandatory expense. On average, corrals or pens around the heart of the homestead where the goats are kept and fed will generally consist of at least woven wire fencing with metal posts. Some homesteads like to make their corrals/pens of wood or welded pipe for better structural integrity, and then add the woven wire fencing to make the containment goat-proof. Pasture fencing is normally constructed with woven wire and metal posts.

Wooden Fence Cost per Foot

Split rail wooden fences will cost $8–$12 per linear foot. Labor will cost ~$10 per linear foot.

Pipe Fencing Cost per Foot

Pipe fencing will cost $10–$15 per linear foot. This is more costly than a wooden corral but will last much longer as well.

Woven Wire Fence Cost per Foot

Regardless of whether wood or pipe is used, woven wire fencing will also be needed. The average cost for a woven wire fence with metal posts will depend on the type of mesh, but on average will run about $1.50–$1.90 per linear foot. This type of fencing is commonly used for pasture fencing.

Electric Fencing

Electric netting fencing for livestock generally runs about $1 per linear foot. This type of fencing can be a good choice for pasture divisions when goats are moved on a monthly or bimonthly basis.

Vet Care

It is not uncommon to require a veterinarian's service occasionally. You might need treatment for a disease or to assist with AI. Therefore, it may be wise to include this as a possible expense item in your annual budget. Medicine and vet care can range from $10 to $50 per month.

Start-Up Expense Summary Charts

The following expense summaries are designed to provide a simple example for demonstrating start-up expenses. The following charts will be based on initial purchase of a 150-lb. milk goat for a homestead with 1 acre of pasture. Feeding, milking, and milk processing equipment are included. The corral will be 60 ft. x 40 ft, with two 20 ft. x 20 ft. internal pens. An enclosed shed (800 square ft.) will be used for shelter and milking. The 1-acre pasture will be enclosed with woven wire fencing.

Initial Livestock Investment Expense Chart

Item	Cost
150-lb. doe	$500

Infrastructure and Equipment Expense Chart

Item	Cost
Woven wire fence 60 ft. x 40 ft.	200 ft. @ $1.75/ft. = $350
Two interior pens 20 ft. x 20 ft.	60 ft. @ $1.75/ft. = $105
One-acre perimeter fenced with woven wire fence & metal posts	836 @ $1.75/ft. = $1463
Milking shed (800 sq. ft.)	@ $15/sq. ft. = $12,000
Shipping container for feed	$2000
Feed trough (6 ft.)	$180
Feed bunk (5 ft.)	$400
Water trough (40 gallons)	$60
Total	**$16,558**

Milk Processing Equipment

Item	Cost
Pasteurizer	$750
Cream separator	$150
Butter churn	$350
Cheese press	$150
Refrigerator	$350
Total	**$1750**

Operating Expenses Summary Chart

The operating expenses are based on milking and maintenance per 150-lb. milk goat for a homestead with 1 acre of pasture.

Feed Costs

Item	Daily Cost	Monthly Cost
Grass hay	$.70	$21.00
Grain	$.78	$23.40
Minerals	$.35	$10.60
Total	**$1.83**	**$55.00**

Animal Care

Item	Annual Cost
Bedding	$190
Breeding cost	$50
Vet care	$120
Total	**$360**

Benefits and Revenue

By having 1 or more milk goats, it will be easier to arrange for the amount of milk to equal your family's needs than it would be with a milk cow. The benefits are mostly that of producing your own milk and milk products, being able to sell unneeded offspring, and the side benefit of family involvement.

Milk Value

Many of the health benefits of having fresh milk for your family can be immeasurable. One of the more quantifiable benefits is perhaps the monetary savings, but this is largely governed by the amount of milk and milk products your family consumes.

Even though a goat will produce less milk than a cow, the value of goat's milk per gallon is more. The price of goat's milk in most areas of the US is $18–$24 per gallon (2024 prices). If your doe produces 1 gallon per day, that is (on average) a $20 value in 2024 prices.

Milk Products

Milk products include mostly butter and cheese. The by-product of making butter is buttermilk. The by-product of making cheese is whey. All of these products and by-products can be used by your family or sold if you have a local demand.

Butter

One gallon of milk will usually yield 1–1.5 pints of cream. The cream can be churned to make approximately ⅓ to ½ lb. of butter.

Cheese

Ten pounds of milk (1.2 gallons) can yield approximately 1 pound of hard cheese or 1.25 pounds of soft cheese (i.e., mozzarella, ricotta, cottage cheese, etc.).

By-Products

The by-products from milk, such as buttermilk and whey, can be consumed, sold locally, or used as a hog feed supplement.

Buttermilk

In the process of making butter, about 25% of the weight of milk will end up as buttermilk. If 1 gallon of milk will produce ½ pound of butter, the by-product will be about 2 pounds or 1 quart of buttermilk.

Whey

In the process of making cheese, 10 pounds or 1.2 gallons of milk will produce 1 pound of cheese. Therefore, about 9 pounds, or a little over 1 gallon of whey, will result as the by-product.

Kid Revenue

If your doe is able to produce 2 kids per year, you can sell the kids for a profit. You should try to keep the kids on mother's milk for at least 6–8 weeks. After weaning and allowing an additional month or two to gain weight, you could sell the kids at an age of 3–4 months.

A kid that is 3–4 months old should weigh approximately 45–50 lb. On average, market prices range from $210–$290 per cwt. For a 50 lb. kid, this would be $145 (50 x 2.9).

For meat, one 50-lb. kid will yield about 27 lb. of meat. At $7 per lb., 27 lb. of meat has a value of roughly $189 per kid. Note that meat prices per lb. can be higher depending on local markets.

Cost/Benefit

The purpose for this analysis is to provide a way to examine the cost versus the benefit of owning a milk goat. By now, you have been able to see that owning and operating a milk goat operation is much more feasible than having a milk cow.

Savings/Loss

The following chart compares the feed and care cost to the milk yield value. The benefit-to-cost ratio will use annual figures in the comparison.

Please note that milk goat purchase cost and infrastructure cost are not included. The same infrastructure can be used for other livestock options.

The savings/loss figure reflects a savings from having to purchase your milk and milk products (*using goat milk prices*) but isn't taking into account the labor involved in milking the goats and processing the milk.

Item	Annual Expense (1 Doe)	Annual Value or Revenue (1 Doe/2 Kids)	Annual Savings/ Loss
Feed & misc.	$660		
Animal care	$360		
Total expense	**$1020**		
Milk value (288 days/yr.)		$5760	
Kid revenue (meat)		$378	
Total savings/revenue		**$6138**	
Net savings/loss			**+ $5118**

- Based on the figures above, the benefit-to-cost ratio is 6:1.
- Note: The milk value is based on goat milk prices.

Picture Gallery

Nigerian Dwarf

Alpine

Saanen

Nubian

La Mancha

CHAPTER 3

Utility Goats

Utility goats can provide a variety of benefits, such as meat, brush control, and fiber. There are a number of breeds that can be used for these purposes.

Almost all goat breeds can be used for meat and brush control, but only a few for fiber. This chapter will list the most common breeds used primarily for meat and brush control, and those used primarily for fiber.

Basic requirements regarding needed space and shelter will be analyzed as well as daily feed costs. This chapter will conclude with a cost/benefit analysis in order to assist you in the decision-making process.

Utility Goats #1
Goats for Meat and Brush Control

The most common breeds of goats used for meat and/or brush control are Boer, Kiko, Spanish, Myotonic (Tennessee), Savanna, and Nubian. These breeds are typically used for brush control and also for producing kids for profit or for meat.

Goats are great for brush control because goats are browsers—not grazers. They prefer to eat browse (woody material) and forbs (weeds); thus, all woody or invasive plant material is considered to be brush.

Using goats for brush control is common in areas where brush is a problem. Goats can clean up an area fairly quickly, depending, of course, on the number of goats used and the amount of brush to be removed.

Boer

The Boer goat is one of the more popular meat goat breeds in the US. It originates from South Africa and is known for its rapid growth rate, high meat yield, and good browsing capabilities.

Kiko

Kiko goats come from New Zealand and have excelled in popularity in the US for their meat production qualities, hardiness, and ability to thrive in various environments.

Spanish

Spanish goats are a diverse breed and often used for brush control and meat. They are hardy, adaptable, and known for their excellent foraging ability.

Myotonic

The Myotonic goat is also known as the Tennessee fainting goat. These goats are often used for meat due to their more muscular form.

Savanna

The Savanna goat is a South African breed and is commonly used for meat production. They are known for their adaptability to harsh climates and resistance to diseases.

Nubian

Nubian goats are primarily used for dairy but are also known for their large size and meat quality.

Goat Selection Considerations

The best approach for choosing goats that will be used for meat and brush control is to focus on a mixture of breeds, sizes, sexes, and try to avoid individuals with horns.

Breed Mix

It is a good idea to have your herd consist of a mixture of breeds because each breed varies in size. This size variation allows for brush control to be more complete and uniform. The tall goats will be able to reach high branches, while the shorter breeds will focus more on the shorter plants.

Another benefit for having a variety of breeds is plant selection differences. Some breeds will prefer various plants and reject others, while a different breed may prefer what other breeds have rejected.

Age, Sex, and Horns

Another helpful variation to have in your herd is a mixture of ages. The younger, shorter goats will browse on shorter plants, and the older, more mature goats will be able to reach higher plants. Another advantage is that younger goats will tend to have plant preferences that can change as they get older.

When it comes to sex, the best mix is does and wethers (castrated males). The wethers will not be preoccupied with the females as a buck would be. Bucks, on the other hand, tend to be aggressive and cause more conflicts. Bucks are also smelly—they will urinate on themselves when around females in heat.

It would be best to have the herd consist mostly of does in order to achieve the most revenue from kid sales or meat. The more the better.

Goats with horns should be avoided if possible. The horns can get tangled in the brush, and they also can get caught in the woven wire or electric fencing.

Diet Management

Goats are browsers, not grazers. They will include some grass in their diet, but primarily they prefer to browse on woody material like brush and forbs (weeds). This is why goats are good at brush control.

Being browsers by nature, they will eat a wide variety of plants—fortunately, the kind of plants that are considered unwanted or invasive species.

Types of Plants

In areas where brush is prevalent, there will generally be a mixture of browse and forbs. Browse consists of small branches, tender shoots, and leaves of woody shrubs, trees, briars, and vines. These are typically medium-to-high reaching plants.

Forbs include stems and leaves of weeds and broad-leaf plants. These are typically low-to-the-ground level plants.

Poisonous Plant Precautions

Some plants are considered poisonous, and extra precautions should be taken to prevent the goats from eating these types of plants. One precaution method is to walk through the area to be browsed and eliminate the dangerous plants before the goats enter. Another way would be to fence off dangerous areas.

Poisonous plants include azaleas, rhododendron, hemlock, bracken ferns, nightshade, rhubarb, larkspur, some varieties of oak leaves, most stone fruit leaves, and some landscaping plants, such as laurels, oleander, and lupines.

Symptoms or toxic reactions include vomiting, frothing at the mouth, diarrhea, and staggering while attempting to walk. If you see these symptoms, you should remove the goats from that area.

Treatment includes hydration with plenty of cool, clean water and administering activated charcoal paste or baking soda (3 tablespoons per quart of water). If the symptoms persist, you may want to consult a veterinarian.

Nutritional Supplements

Depending solely on browse may not be enough to keep your goats healthy. They should be getting plenty of forage but may also need additional nutrition with a grain and mineral supplement. For a brush goat herd, the basic need for a grain supplement can be met by feeding roughly 1 cup per day for adults and ½ cup per day for kids.

Goat Management

Managing your brush control goats in the field includes fencing, shelter, and water. Every time your goats are moved, you will need to provide these three essentials.

Fencing

As mentioned previously, all goats are notorious for trying to escape containment. They will test your fencing by trying to push the fence over by leaning on it, or look for a way to go over, under, or through an opening.

Woven wire fencing works well for long-term fencing but can be too expensive and laborious for short-term brush control. With that in mind, electric fencing is a popular solution. One type of electric fencing that works well is electric sheep and goat netting.

This type of electric fence can be both installed and later removed fairly easily and quickly. This type of fencing runs about $1 per linear foot.

Shelter

Goats are herd animals. Even though they may spread out during the day while browsing, they like to congregate at night and bed down together.

It is good to provide shelter, if at all possible, for protection at night—especially from harsh weather and possible predators. If the goat herd is small, and remaining on your homestead, you should have a permanent shelter available for them.

If your brush control herd is being transported and hired out for brush control at distant sites, you might want to consider the use of a mobile shed on wheels. Of course, this would only be feasible with a small herd and if roads are accessible.

Another option for distant field use is a hoop house type shelter. The frame can be aluminum pipe or PVC, and the top covering can be a canvas tarp. This can be mobile and easily assembled and disassembled in the field.

On average, about 12–15 sq. ft. per goat is adequate for indoor space. In your permanent shelter, remember that you can build pallet-sized shelves on the walls of the structure, on which the goats like to hop up onto and sleep. This can help provide extra square footage for bedding space.

Water

You should have clean, cool water available on your homestead for all of your livestock. If the brush goats are kept on the homestead, they should be allowed to return to the shelter with adequate water and supplemental feed at night.

If the brush goat herd is being used remotely, you will need to arrange for water to be hauled to the area where the goats are browsing and kept overnight. This can be arranged by having a water tank on a trailer and having portable water troughs for water availability. If your herd is relatively large, you may need more than one water trough to prevent crowding.

On average, goats need 2–3 gallons of water daily. If the goats have plenty of lush green browse to eat, they may get by with 1–2 gallons per day. Lactating does may require at least 3 gallons of water daily.

Brush Control Management

The strategy of using goats for brush control is not a new idea. In fact, it is growing in popularity as a means to reduce fire hazards, because the price of mechanical alternatives has been increasing.

The use of goats for brush control will require careful planning in order to get the most gain from your efforts. To be properly managed, you will need to know how many goats to use per acre, and the correct duration, in order for your operation to be efficient.

Brush Control Benefits

It is possible to clear brush mechanically, but the process requires a lot of labor. The brush has to be cut or mowed, then removed to eliminate the fire hazard of dry woody material. By using goats, the brush is broadly reduced with no removal necessary.

When an area has been browsed and cleared, the browse line will be high, and the only plants remaining, for the most part, will be grass. Goat droppings will also serve as fertilizer for the remaining trees and grass, which can assist in erosion control.

Goats per Acre

How many goats will be necessary to clear an area and how long will it take are common questions. The answers are dependent on how thick the brush is, the types of plants needing to be cleared, and the herd makeup (i.e., breeds, sexes, ages, etc.).

On average, you can use the following formula to determine how many goats will be needed per acre.

- 1 goat can clear .1 acre in 4 weeks, or
- 10 goats can clear 1 acre in 4 weeks.

- 1 goat can clear .025 acres in 1 week, or
- 40 goats can clear 1 acre in 1 week.

Goat Movement

You will need to be prepared to move your goat herd after an area has been adequately cleared of brush. You should already have the next area planned and ready for the herd to be moved to in order to prevent the common problems associated with waiting too long.

Timing

It takes time to graze down an area. As discussed earlier, it depends on the size of the herd, the number of acres, and amount of brush to be cleared. Goats will tend to focus on preferred plants first and will only be willing to eat the less-preferred plants when those plants are the only thing available.

Brush removal depends on the number of woody species present, but the removal percentage is somewhere between 50% and 90%. You must have in mind what it should look like, when the target removal goal has been reached.

When the target removal level has been reached, or is close, it is time to move the herd to the next area. If you wait too long, the goats will begin to damage desirable trees by eating bark; be more inclined to eat dangerous, poisonous plants; or be more aggressive in trying to escape.

Repeat Grazing

Brush control can serve two purposes: (1) as a food source and (2) as a viable method of brush removal. Your ultimate goal in brush removal is to eliminate undesirable brush and weeds. This will happen over time, but it is a slow process that will require multiple treatments over a period of several years.

The timing of the treatments is also important. For the treatment to have the desired effect, the browsing should take place in the spring and summer months. In the fall and winter months, shrubs and trees will be going dormant, and the "pruning" by way of browsing will not deter the next season growth cycle.

Goat Health

On average, goats are fairly hardy and don't require a lot of maintenance, nevertheless, there are some diseases and pathogens that can become problematic. For a list of possible maladies and treatments, you can refer to Health Precautions/ Maintenance (for Milk Goats) in Chapter 2, Part 1.

Goat Protection

Goats, like most small farm animals, can be subject to predation. For that reason, you will want to monitor your herds on a regular basis and take the necessary precautions against the possibility of predators in your area.

Monitoring

Monitoring includes both an on-site visual observation of their health and well-being and being vigilant for the possible danger of predators. This can be done by checking the perimeter for tracks, checking for signs of injured goats, and counting to make sure the numbers are what they should be.

Being vigilant can include trail cameras stationed around the nighttime bedding-down area. These cameras should be checked daily in order to assess the threats and deal with them as soon as possible.

Predator Control

It is not uncommon for homesteads to experience predators that threaten the livestock, especially those that aren't protected at night by way of shelters. If your goat herd is away from your home base in a remote location, you should have some sort of plan for predator control.

One common method for protecting a goat herd is to use a guard dog. Livestock guardian dogs are among the most popular and effective choices for protecting goats. Breeds such as the Great Pyrenees, Anatolian Shepherd (also known as the Kangal Shepherd), and Maremma Sheepdog are well-known for their ability to deter predators and bond with their goat-guarding responsibilities.

Meat Production

Meat production is the other benefit that comes with owning a goat herd. The percentage of meat you can get from a goat when butchering can vary based on several factors, including the age, size, and overall condition of the goat, as well as the specific cuts of meat you're interested in. However, the following chart provides a general range of meat yield percentages for the typical brush/meat goat breeds.

Criteria	Boer	Kiko	Spanish	Myotonic	Savanna	Nubian
Average wt. doe	200–300	130–180	80–150	60–100	130–200	135–175
Average wt. buck	250–380	180–250	100–180	75–160	150–250	175–250
% of live wt. in meat	40–50%	35–40%	30–40%	25–35%	40–50%	25–35%
Average lb. of meat/ adult	112	72	45	30	67	52

• **Note:** These are average figures for adult goats.

Expenses

The basic expenses for purchase costs, feeding, feeding equipment, water equipment, containment, shelter, and vet care are going to be similar to dairy goats. The details can be viewed in the Expenses section (for Milk Goats) in Chapter 2, Part 1.

Revenue

Revenue can possibly come from brush removal, if you are hiring out your herd for that purpose. However, most often the benefit is realized on your own property, and the revenue, therefore, is from kid sales and/or meat sales.

Brush Control

The value for brush removal and weed abatement is generally valued at a dollar figure per acre. If you were hiring out your herd for this type of service, the typical charge is $400–$1000 per acre. This figure varies depending on the size of the site, terrain, amount of brush, accessibility, etc.

Kids for Sale

On average, your does will have twins once a year. The typical weaning age is about 3–4 months. A kid that is 3–4 months old should weigh approximately 45–50 lb. On average, market prices range from $210–$290 per cwt.

Kids for Meat

Kids can also be raised for meat. One 50-lb. kid will yield roughly 27 lb. of meat. At $7 per lb., 27 lb. of meat has a value of roughly $189 per kid. Note that meat prices per lb. can be higher depending on local markets.

Feed Conversion Ratio

The feed conversion ratio (FCR) is the measure of livestock production efficiency. It is the comparison of feed intake to weight gained over a period of time.

Typically, the kids from your goat herd are weaned at 3–4 months old and weigh, on average, 45–50 lb. For the first several weeks, the kids will be consuming mostly milk.

If a kid eats, on average, 1.5 lb. of forage per day and .5 lb. of grain per day, during months 2–4 (90 days), the overall feed intake during this 90-day period is 180 lb. The gain over this period is 30 lb. (total 50 lb. minus 20 lb. birth weight).

Criteria	Total Forage Intake (Lb.)	Total Grain Intake (Lb.)	Weight Gain (Lb.)
Pounds per day	1.5	.5	
Total lb. months 2–4 (90 days)	135	45	30

The gain of over 30 lb. over a 90-day period = .3-lb. gain per day. The feed conversion ratio (FCR) for the above metrics would be 180-lb. feed intake and 30-lb. gain in 90 days, which yields an FCR of 6:1.

Cost/Benefit

The benefit of having a goat herd used for brush control and meat is variable. It largely depends on how much the feed cost is offset by browsing. If the herd spends a lot of the year browsing, the revenue from kid sales and meat can potentially yield a profit. If the herd is hired out to neighbors for brush control, profits can be realized from both brush control and from the kid crop.

However, if the herd spends little to no time browsing, and must rely on supplemental feeding, the overfeed cost may be higher than the revenue generated from the kid crop.

Utility Goats #2
Fiber Goats

Goats used for fiber basically fall into two categories: Angora and Cashmere. There is only one Angora goat breed, but Cashmere is not a breed—it is a type of goat that produces cashmere as a certain percentage of its wool.

Fiber goats, unlike sheep, are sheared twice per year. The cashmere fiber is known for its softness, light weight, and warmth. The fiber is hollow and thus insulating, which allows it to be significantly warmer than sheep's wool. The hollow fiber has the ability to "breathe" and to also wick moisture from the skin, which allows a cashmere garment to be comfortable, even in warmer weather.

Goat Breeds

The following breeds are used for fiber. The only goat that produces mohair is the Angora goat; the remainder of the breeds listed are known for their cashmere wool.

Angora

Angoras originate from Asia Minor and Turkey and are bred for their highly sought-after fiber called mohair. The fleece is similar to that of a sheep and is considered valuable due to its softness and warmth. The Angora goat is a large goat and can be of a variety of colors (white, brown, black, or red).

Hexi Cashmere

The Hexi Cashmere goat breed originates from the desert and semi-desert regions of China. They have the ability to adapt to harsh environments and produce fine, soft undercoats of cashmere. They are considered to be a medium-sized goat, and their colors are typically white, brown, or black.

Changthangi

The Changthangi goats originate from the high-altitude regions of Tibet and Nepal. The Changthangi wool is an ultra-fine cashmere fiber. This fiber is called pashmina, which is one of the world's finest cashmere fibers. The Changthangi goats are considered a small-to-medium-sized goat and are generally white in color.

Nigora

The Nigora goat is an American hybrid cross between Angora goats and Nigerian Dwarf goats. The Nigora goats' wool fiber has three classifications: Class A (mohair fiber), Class C (cashmere fiber), and Class B (a cashgora fiber, which is a cross between angora and cashmere fibers). The Nigora is a small-to-medium-sized goat, and the color can be varied.

Pygora

The Pygora goat is a hybrid cross between Angora goats and Pygmy goats. The Pygora goats' wool fiber (like the Nigora goat) has three classifications: Class A (high quality mohair fiber), Class C (cashmere fiber), and Class B (a cashgora fiber, which is a cross between angora and cashmere fibers). The Pygora is a small-to-medium-sized goat and has a variety of colors, such as white, black, brown, and gray.

Breed Comparison Chart

Criteria	Angora	Hexi Cashmere	Changthangi	Nigora	Pygora
Origin	Turkey	China	Tibet, Nepal	US	US
Height (in.) doe/buck	36	24–28	22–26	20–27	18–26
Weight (lb.) doe/buck	110/224	80/100	57/68	60/100	75/95
Fiber type	Mohair	Cashmere	Cashmere	Cashmere	Cashmere
Wool per year	8–16 lb.	4–8 oz.	4–8 oz.	1–3 oz.	4–10 oz.

Other Breeds

There are other Cashmere goat breeds that are less common in the US. These breeds include Altai Mountain, Australian Cashmere, Don, Kaghani, Liaoning Cashmere, Licheng Daqing, Luliang Black, Tibetan Plateau, Wuzhumuqin, Zalaa Jinst White, Zalawadi, and Zhongwei.

Goat Maintenance

Breeding

It is best management practice to arrange for one set of offspring per year. Like other goat breeds, fiber goats typically have a gestation period of 150 days or roughly 5 months. Fiber goats typically breed in the early fall, about October, allowing the does to kid during February and March.

Like other goat breeds, fiber goats are also able to have twins, and this likelihood is sometimes enhanced by increasing the grain rations in the fall during breeding time.

Unlike dairy and meat/brush goats, artificial insemination is not normally used, and bucks are either allowed to run with the does during the breeding season or turned in with does when they come into heat.

Feed and Water

The diet for fiber goats is similar to other goat breeds. They are browsers, and their diet usually consists of a mixture of browse and/or grass hay and grain. Like other goat breeds, their daily diet can consist of roughly 5 lb. of forage and 4 cups or 2 lb. of grain per day.

Clean, cool water should be available during the day. On average, each goat will consume about 2–3 gallons per day.

Shelter

Even though fiber goats can withstand cold, windy, or wet conditions, they prefer, like other goat breeds, to have shelter to protect them from harsh weather.

You should arrange the shelter to provide adequate space. Each goat should have, on average, 16–20 sq. ft. to ensure they have adequate space.

The fiber goats should have clean, dry bedding to keep them healthy and avoid disease. Bedding material can be sawdust or straw.

Healthcare

In order to keep your fiber goats healthy, it is recommended to maintain a careful routine of monitoring as well as appropriate and timely preventative maintenance treatments and vaccinations.

Some of the more common health issues that can occur include coccidiosis, tapeworms, pregnancy toxemia, foot rot, and sore mouth.

Fiber Management

The following description is based on how to process mohair. The harvested weight of the mohair per animal is significantly higher than that of cashmere.

Pre-Shearing

Before the shearing begins, you should remove all foreign particles from the mohair, such as straw, dirt, matted droppings, etc. For the shearing process, you can place the goat on a milking stand, so they can't move while shearing.

Shearing

Each animal is sheared twice per year. This is usually accomplished by using electric clippers. The resulting fleeces are rolled up and placed in bags.

Mohair Processing

Scouring

The mohair is cleaned by a process called scouring. You would have several tubs of warm, soapy water for washing and clear water for rinsing. The mohair is then pressed and blown dry. The resulting clean mohair fleece is usually 20% lighter after the lanolin and dirt has been removed by the scouring process.

Stock Dyeing

Once dry, the mohair can be dyed. When dyed after scouring, it is referred to as stock dyed.

Carding

Carding helps to remove any remaining pieces of straw and straightens the fibers so they will lie in the same direction. The mohair fibers are brushed through a series of metal or wire brushes that separate and align the fibers into a thin thread. The thread is gathered into narrow strips and joined to form the roving.

At the roving stage, the fiber is typically in a long and narrow form, making it easier for spinners to work with. Roving is a crucial intermediate step in turning raw fibers into usable yarn. It ensures that the fibers are well-prepared, aligned, and ready for the spinning process, leading to more consistent and high-quality yarn production.

Drafting

After carding, the fibers are drawn out into a thin and even bundle. This process is called drafting, and it further aligns the fibers while making them more manageable for spinning.

Twisting and Spinning

The drafted fiber bundle is twisted together to form a continuous strand. This twisting process not only holds the fibers together but also gives the yarn additional strength. The thickness of the yarn depends on how much fiber is drafted and twisted together.

Yarn Dyeing

The yarn can be dyed after spinning, which is referred to as yarn dyed. This process increases the value of the yarn.

Weaving

Woven fabrics are made on looms by interlacing at least two sets of yarn, either woolen or worsted, at right angles to each other. The lengthwise yarn is the warp. Threads running crosswise in the loom are called weft or filling. As warp thread passes through the loom, it is raised and lowered by a wire eyelet through which it is threaded. Filling thread is passed through the openings created in the warp to form the woven fabric.

Finishing

As the fabric comes from the loom, it is inspected for defects. The fabric can then be napped by a metal brushing process or sheared to give a smooth uniform appearance. Various chemical finishes can be applied to obtain such advantages as mothproofing, stain resistance, and washability.

Fiber Sales

On average, Cashmere goats produce 6 oz. of wool, and it sells for $8–$12 per oz. That is roughly a $60 value per goat for premilled cashmere. The Angora goat produces roughly 12 lb. of mohair, on average, and it sells for $7–$12 per pound. That is an average of $120 per goat for premilled mohair.

Homesteaders wanting to profit from fiber most often choose to go with the Angora goat because of the higher production per goat. In addition, most choose to sell their mohair in the dyed yarn stage in order to boost the profitability.

For mohair, dyed yarn can bring $6 per oz. or $96 per pound. The following graph shows the end profit per goat after expenses.

Item	Expenses	Income	Profit
Annual feed	$659		
Bedding, vet, etc.	$260		
Vet and misc.	$120		
Kid sales (2)		$400	
Mohair processing 12 lb. @ $10/lb.	$120		
Dyed yarn ($6/oz.)		$1152	
Total	**$1159**	**$1552**	
Profit			**$393**

If you process and dye the mohair, and sell the kid crop each year, the overall endeavor can be a profitable enterprise. However, if the mohair is only sold by the pound after shearing, the total revenue for mohair plus kid sales would be about $620. This, on average, would end up being a loss of about $419 after expenses.

Picture Gallery

Meat and Brush Goats

Boer

Kiko

Spanish

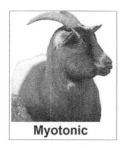
Myotonic

Savanna

Nubian

Fiber Goats

Angora

Hexi Cashmere

Changthangi

Nigora

Pygora

CHAPTER 4

Sheep

Sheep can be great fun on a homestead. Sheep are gentle and docile, and their lambs are adorable, making for a great family experience—especially children. If your goal is to have a sweet animal experience, and cost-effectiveness isn't a priority, then you can't go wrong with sheep. It is possible to make a profit raising sheep, but that will require a reasonable amount of acreage for grazing.

The main reason for sheep's popularity is their dual purpose. This chapter will list the breeds used mostly for wool, and those used for both wool and meat.

Basic requirements regarding needed space and shelter will be analyzed as well as daily feed costs, animal management, wool management, pasture management, predator control, and profitability assessment. This chapter will conclude with a profit/loss analysis in order to assist you in the decision-making process.

Breeds

There are many sheep breeds used in the US. The majority of those breeds fall within the categories of fine, medium, and coarse wool breeds. Many of these breeds are used for meat as well as wool. Listed below are a few of the more commonly used breeds from the fine and medium wool categories.

Fine Wool

Delaine-Merino

The Delaine-Merino sheep breed is a type of domestic sheep known for its high-quality wool production. It is a result of crossbreeding between the Delaine sheep and the Merino sheep. The cross allows for the breed to produce a soft, fine, superior wool and a high-quality meat yield.

Rambouillet

The Rambouillet breed was developed from the Spanish Merino sheep and is raised for both wool and meat production. Rambouillet sheep are known for their fine wool and are prized for their excellent fleece quality.

Debouillet

Debouillet sheep are a breed derived from Rambouillet and Delaine-Merino crosses. They also produce high-quality fine wool and are well adapted to various climates, making them suitable for wool production in different regions.

Medium Wool

Southdown

Southdown sheep are a smaller breed known for their meat quality. They have a distinctive appearance with a round body and a characteristic "dished" face. They are popular for meat production due to their tender and flavorful meat.

Hampshire

The Hampshire sheep are recognized for their impressive meat production. They have a distinctive black face and legs, and their meat is high quality and flavorful. Hampshire sheep are widely raised for meat in various farming systems.

Suffolk

The Suffolk sheep are another meat breed, valued for their rapid growth and muscular build. They have a distinctive black face and legs, and their meat is lean and tasty. Suffolk rams are often used for crossbreeding to improve meat characteristics in commercial flocks. A common cross is with the Rambouillet breed that is known for high-quality wool.

Columbia

Columbia sheep are a dual-purpose breed developed in the United States. They are known for both their meat and wool production. They have a white face and legs, and their wool is of good quality. Columbia sheep are adaptable and well-suited for diverse farming environments.

Other Breeds

Other less common sheep breeds found in the US are Targhee, Finnsheep, Corriedale, Montadale, Cheviot, Shropshire, Oxford, Border Leicester, Lincoln, and Romney.

Breed Comparison Chart

Criteria	Delaine-Merino	Ram-bouillet	De-bouillet	South-down	Hamp-shire	Suffolk	Columbia
Ht. (in.) female/ male	28/34	29/33	28/32	24/28	30/34	30/34	28/32
Wt. (lb.) female/ male	175/250	140/200	125/200	120/180	175/225	200/250	160/250
Wool wt. (lb.)	8–15	10–15	9–14	5–8	6–10	6–12	12–18
Meat yield (% live wt.)	40–55	45–55	45–55	40–55	50–60	50–60	45–55

Gestation

The gestation period is the time between conception and lambing. The gestation period for domestic sheep is commonly 147–150 days or roughly 5 months.

Breeding

On average, ewes will come into heat roughly every 17 days, and it will last 24–36 hours. Even though a gestation period of 5 months would allow for two pregnancies per year, this is not the normal practice. The general practice is to arrange for the ewes to be bred in the fall and lamb in the spring.

In order to control and manage the timing of lambing, the rams are generally kept separate from the ewes except during the fall breeding season. A single ram can usually service up to 40 ewes.

If your operation is small, and you don't want the expense of keeping a ram year-round for fall service, other options may be available. You may be able to borrow a neighbor's ram, or you may be able to arrange for your ewes to be artificially inseminated.

Feeding

Sheep are grazers and prefer to eat mostly grass. The ideal situation would be to have pasture. If this isn't possible, or not available during certain months out of the year, alfalfa hay or pellets will need to be provided.

For general maintenance, the daily minimum forage intake is about 3% of body weight, increasing to 3.5% during late gestation and early lactation. For a 140-lb. ewe, the daily forage intake would be 4–5 lb. of forage.

Ewes need additional sources of protein, such as soybean or cottonseed meal, during gestation and lactation—especially if the ewe is going to have twins. About 70% of fetus growth occurs during the last 4–6 weeks of gestation. Underfeeding during this time can lead to toxemia (ketosis).

Sheep are generally shorn before lambing. After being sheared, ewes will require more feed to compensate for heat loss.

The daily amount for grain supplement during gestation and lactation would be about 1.5% of body weight. For a 140-lb. ewe, the daily grain supplement would be roughly 2 lb.

Feeding Equipment

Feeding equipment will normally consist of a feeding trough for grain supplement and a feed bunk for hay.

Feed Troughs

Feed troughs are generally made of metal or wood when in a pasture. Plastic versions can be used, if preferred, while in a pen. These troughs can be semicircular or square in shape. They can be handmade or purchased.

Feed Bunks

Feed bunks are generally used for feeding hay. To avoid waste, the feed bunk should be sized specifically for sheep.

Feed bunks are designed to hold fodder such as hay in the upper part, with a trough underneath to catch the hay as it is pulled out from the bunk in the feeding process. Feed bunks can be made of wooden boards or metal.

For wooden versions, the frame can be constructed with 2" x 4"s. The upper bunk slats are generally constructed using 1" x 4"s, about 4 ft. long and 3–4 inches apart—wide enough for the sheep to get their nose through. The lower trough can also be constructed with 1" x 4"s or wider lumber.

For the metal version, the frame can be constructed by welded metal pipe. The slats can consist of welded rebar, and the lower trough can be constructed by cutting 16-gallon barrels in half, and the edges covered with angle iron welded into place.

Both wooden and metal feed bunks are becoming more available for purchase at your local feed store or online.

Feed Storage

Feed storage structures are used to store hay and grain. These structures can be constructed with wood or metal. Even shipping containers make excellent feed storage structures, which can be relatively inexpensive.

Watering

On average, sheep will drink 3–5 gallons of water per day. This volume depends on the type of feed being consumed, age, ambient temperature, and if they are lactating.

The common watering equipment for sheep is galvanized water troughs or tanks. Heavy plastic water troughs can also be used.

Lambing

Lambing season can be a fun time, especially if you have a small herd. There is nothing more precious than to see a field full of lambs running and playing. It is not uncommon for ewes to have twins, so a management plan should be carefully monitored in order to prevent losses. With that in mind, it is important to be prepared when lambing season is approaching. Preparations include increasing daily feed rations for pregnant ewes, making sure the lambing facilities are ready, and ensuring you have the necessary supplies on hand.

Space Requirements

It is possible for ewes to lamb in a pasture, if you have a large herd and large pastures, but to reduce lambing losses, and reduce the risk of predation, most homesteaders choose to have a shelter to use as a lambing facility.

The lambing facility is normally a barn that includes small lambing pens inside for ewes giving birth. The pens are typically

4 ft. x 6 ft. in size or 20–25 sq. ft. The pens should be clean and have fresh bedding of straw.

After several days, when the ewes and lambs have had a chance to bond and gather their strength, they can be allowed back into the pasture during the day.

Orphans

Occasionally, you may have a ewe that has twins, but only accepts one, leaving the rejected lamb an orphan. When this happens, you will need to have milk bottles with nipples to feed these lambs twice a day. These supplies can be purchased at a local feed store as well as a powdered milk that contains the necessary nutrients.

On occasion, you may have a situation where you have an orphan from 1 ewe, and another ewe's lamb died. There are two methods you can use to encourage the ewe that lost her lamb to accept the orphan as her own.

One method is to skin the dead lamb and tie the pelt on the orphan. The other method is to spray deodorant into the nostrils of the ewe that lost her lamb and then also spray the orphan lamb with the same deodorant. Both of these methods need to be applied on the day the dead lamb died.

Docking

It is a normal practice to remove the tail, which is called docking. The most common way of doing this is by using rubber bands made for this purpose. These rubber bands are small, thick bands about the size of a nickel. They are applied with the use of a plier-type tool called a ring extender. The rubber band, once applied, will cut off the blood supply, and in a few weeks, the tail will simply fall off. It is an easy and painless process.

Wool Management

The sheep will need to be sheared once per year, and this is normally done in the spring, just before lambing. With the proper equipment, using the proper technique, it can be a relatively easy process. You will need pens, chutes, electric shears, and wool bags.

Shearing Process

The shearing process is relatively simple, but the correct equipment will make things go much faster. If you plan on doing your own shearing, electric clippers will be a must.

Separation

You should have one pen large enough to hold all of the sheep, with a chute attached to one side. The sheep should be forced into the chute, one at a time, until the chute is full. One person will be needed to keep the sheep in the chute moving forward.

At the shearing end of the chute, the shearer will extract 1 sheep at a time, and after the sheep has been sheared, the shorn sheep should be ushered into a separate pen for the finished sheep.

Shearing

Sheep shearing is a skill that needs to be acquired. It is possible to find professionals to shear your sheep, but it may be difficult to find someone, unless you have a large herd to make it worthwhile to them. Most charge $3–$5 per head.

If you are doing it yourself and don't have the experience, it is not uncommon to use a shearing stand, similar to a goat milking stand, that has a stanchion on one end to hold the sheep in place. A professional will shear the sheep without a stand, as described below.

The shearer will extract a sheep from the chute, and the sheep is typically pulled up between the shearer's legs, with the sheep sitting on their bottom, and the sheep's head under the shearer's arms. Having the sheep in a sitting position helps to keep them inactive during the shearing process.

Starting on the stomach, the shearer will begin to shear with a downward motion. The wool is clipped close to the skin, and the fleece will hold together as it is being removed. Moving up and down the body of the sheep, the shearer will work their way around the legs and slowly move their way around the side of the sheep, ending up where they started. When finished, the shorn sheep will be released into the holding pen.

Skirting

After shearing, the fleece needs to be cleaned, folded, rolled, and tied. To skirt the fleece. place the fleece flesh side down, so the dirty side of the fleece is facing up. Next, remove off-color wool, matted wool, imbedded chaff, and any soiled areas.

Next, fold over the sides of the fleece toward the center, and then roll the fleece from one end to the other. The flesh side, or clean side, will face outward, presenting a clean-looking fleece for the buyer. It may or may not be necessary to tie fleeces. It depends on the buyer's preference. For fleeces that are tied, only paper twine should be used. The fleece can then be placed in a wool bag to await more precise packaging later.

Packaging

For small producers, it is a common practice to protect the fleeces by first putting them in clear, plastic bags, and then package the bags in cardboard boxes. Black or brown wool should be bagged and packaged separately from pure white wool and be properly labeled. Different wool grades or classes (i.e., fine, medium, coarse, etc.) should be packaged separately and labeled accordingly.

All low-quality wool, like belly wool, off-color, chaffy, stained, or matted wool should be handled and bagged separately. Properly sorted and labeled wool will yield a higher price.

Wool Quality

Wool quality can be determined by a number of factors, such as fiber diameter, crimp, yield, color purity, staple length, and fiber strength.

Fiber diameter is the measurement of the wool fiber thickness and reflects the fineness of the fiber. The diameter is measured in microns, and the lower the micron, the finer the wool. Fiber diameter can vary somewhat throughout the fleece. The higher the variation, the lower the quality. Conversely, fleeces with uniform fiber diameters are considered much higher in quality.

Wool typically has a natural waviness in the fiber called crimp. The more crimp the fiber has, the finer the wool will be.

Yield is a term used to express wool weight after washing (scouring). This is typically measured in pounds, and the amount of weight loss after scouring is called shrinkage. This is largely contingent on how dirty the fleece was before washing.

The value of color is variable. In most markets, pure white wool is preferred because it can be dyed—black and brown fiber cannot. To a much lesser extent, some markets prefer fiber with color, as some spinners and weavers prefer the natural colors.

One of the more common determining factors is staple length. The longer the fiber, the higher the value, because it is easier to spin. Fiber strength is another important fiber characteristic. Strong fibers are able to withstand the harsh demands that fibers are subjected to during wool processing.

Wool Sales

The price of sheep wool is much lower than mohair or cashmere. The main reason is supply. There are over 5 million sheep raised in the US each year. That is a lot of wool.

The two most common avenues to sell your wool are to a commercial broker or through a fiber coop. According to the USDA NASS (US Department of Agriculture, National Agricultural Statistics Service), the US average raw wool price per pound for 2021 and 2022 ranged from $0.40 to $1.70. Even at $1.70 per pound, the value for a 12-lb. fleece was only $20.40.

In order to raise the value of the wool, direct marketing is a more viable option. This, however, requires some form of wool processing.

Wool Processing

Direct marketing involves the promotion and sale of your product directly to the consumer. Marketing your fleeces directly to hand spinners and weavers is the most commonly used approach.

Once the wool has been cleaned, it is then carded. This can be done by hand but is usually done by machine. The carded wool is made into a thread that is gathered into long strips and joined to form a roving. Roving is an important step in turning raw fibers into usable yarn.

The roving thread is then drafted into a bundle. The drafted fiber bundle is then twisted together to form a continuous strand called yarn. For a more detailed description on wool processing, you may refer to Fiber Management (for Utility Goats) in Chapter 3, Part 1.

The spun yarn can then be woven into usable forms, such as woolen garments and blankets. Dyed yarn and woolen garments and blankets can then be sold directly to the end consumer online or at farmers' markets at a much higher price.

Wool Used as Insulation

Sheep's wool is a great insulator and is often used for home wall insulation. Wool can be blown into place between the studs, or it can be inserted like conventional fiberglass batts. The R-value is 3.5–4.5 per inch.

It is effective for temperature stability and sound absorption. Wool is flame resistant, and because wool is a keratin, it is mold and mildew resistant as well. The wool fibers are breathable, allowing the wool to absorb and release moisture.

It is possible for insects, like moths, to become a problem in the wool, but this type of threat can be eliminated with a borax treatment before installing.

Pasture Management

Pasture management is the practice of managing the number of animals placed in a pasture for grazing and the duration of grazing in that pasture. Both numbers and duration can adversely affect the health and productivity of the pasture, if not properly managed. For a full explanation, please refer to Pasture Management (for Milk Cows) in Chapter 1, Part 1.

Grasses

There are numerous species of grass that can be introduced and used for grazing in areas with moderate-to-high rainfall. In areas where rainfall is limited, you must rely on native species.

The most common native grasses in the southwestern US are Blue Grama, Side-oats Grama, and Buffalograss. Grasses native to most other western US states are Bluebunch Wheatgrass, Idaho Fescue, and Big Bluestem. Grasses native to most eastern US states are Little Bluestem, Switchgrass, Indiangrass, Easter Gamagrass, and Wild Rye.

Healthy Plant Growth

Healthy grass plants will use photosynthesis to convert the sun's energy and carbon dioxide into glucose and oxygen. The glucose is used for growth and also stored in the roots for future regrowth in the form of starch or carbohydrates. As the plant is allowed to grow, more and more energy is stored in the roots for future growth.

Overgrazing

Overgrazing basically reverses the healthy growth trend. Continuous grazing causes the plant to continually rely on stored food for regrowth. This will eventually lead to the stored food in the roots to be exhausted, causing the plant to become weak and ultimately die. Nature will replace the grass plant with weeds and other fast-growing undesirables.

Carrying Capacity

The best way to prevent overgrazing is to use a rotational grazing system. First, however, you must determine how many head the pasture can carry without being overused. This is called carrying capacity, which is typically represented in animal unit months (AUM). One AUM is the amount of forage that can sustain one 1000-lb. cow or 5 sheep for 1 month.

There are two common methods for determining carrying capacity: (1) the estimation method and (2) the sampling method. For a more thorough explanation of these methods, please refer to Pasture Management (for Milk Cows) in Chapter 1, Part 1.

The more common method for homesteading is the sampling method. The result will yield an AUM amount. One AUM equals one animal unit month (for one 1000-lb. cow). To convert this AUM number to sheep, divide the AUM number by .2.

Rotational Grazing

Without rotational grazing, the pastures are generally grazed unevenly, meaning some areas are grazed more heavily than others. The main benefit of rotational grazing is to allow the pasture to be grazed more uniformly and to allow the grass time to regrow by having a rest period before being grazed again. This allows the grass to recover and replace the needed energy in its root system.

Let's say you have 4 acres, which is shown to have a carrying capacity of 12 AUMs. This converted to sheep would be $12 \div .2 = 60$ sheep AUMs. That means you can have 60 sheep on 4 acres for 1 month, or you can have 5 sheep on 4 acres for 12 months—or any combination equaling 12 AUMs.

To introduce rotational grazing, the acreage could be divided into pastures. If the choice were to have 5 sheep on 4 acres for 12 months, the acreage could be split into 2, 3, or 4 pastures (i.e., 2 pastures for 6 months each; 3 pastures for 4 months each; or 4 pastures for 3 months each).

Obviously, the more pastures created, the more fencing is required. This can also be achieved with electric fencing, reducing time, labor, and materials. In a general sense, the more pastures you have, the less impact on the soil from grazing you will have, but water availability may govern that decision.

Herding

The more sheep you have, the more challenging it can be to herd them in a certain direction, when it is time to change locations. The reason herding is a challenge is because all sheep are followers—no one wants to be the leader.

One common way to herd your sheep is with a lead animal that will guide them where they need to go. A goat is often used for this purpose, which is commonly called a Judas goat. Another common means to herd your sheep is by using a sheepdog. This, of course, works well, but it warrants having large acreages and large herds to justify the expense.

Replanting

One thing that is common with sheep grazing is that in time, the pasture ground is going to become hard due to the sheep hoof trampling and tamping effect. This phenomenon will reduce the ability for plants to reproduce and remain productive.

The solution is to periodically (every 4 years or so) disc the pasture and reseed. The best seed for sheep is a seed mixture of both native and introduced species. The best mixture for sheep will also include one or more clovers.

Coordinated Grazing and Browsing

For those homesteads having acreage with brush, both goats and sheep can be used in concert. In theory, the goats can be used for brush control, and the sheep can graze on the grass underneath. This allows both the sheep and the goats to work together for profit, if there is enough acreage to warrant the enterprise.

Containment

If you are raising sheep on your homestead, you will need a large pen near the barn and feeding area. If you have acreage for grazing, the pastures will need to be fenced as well.

Livestock Feeding Pen

The large livestock feeding pen for ewes and lambs is normally constructed with woven wire fencing material. Typically, the woven wire fence is approximately 3.5 ft.–4 ft. tall, and the square openings no larger than 4" x 4". The fence will need to be stretched tight, with the metal T posts spaced about 12–15 ft. apart.

Lambing Pens

The large livestock pen should be in close proximity to the barn, so that pregnant ewes can use the lambing pens during the lambing season. Lambing areas are typically arranged inside or under some sort of shelter, like an enclosed shed or barn. Lambing areas can be isolated pens, or they can be more confined stalls.

The size of the stalls can vary, but the typical size is 4 ft. x 6 ft. The back and sides of the stall are normally constructed with ½" to ¾" plywood, 4 ft. tall, with a front gate made with woven wire fencing material, framed with wooden 2" x 4"s.

Pasture Fencing

Fencing for pastures is generally constructed with woven wire fencing and metal posts. The square size of the woven wire mesh should be no larger than 4 inches. The height of the fence should be about 4 ft. tall.

Internal pasture divisions can be constructed with woven wire fencing or by using electric fencing. Electric netting makes the best electric fence for sheep.

Shelter

Any homestead should have appropriate shelters for the various types of livestock. Sheep should have a shed or barn available for bedding down at night—especially during lambing season—to protect them from harsh weather and possible predators.

The most ideal form of shelter is a wooden barn. Other types of shelters that can be used are enclosed sheds. These structures can be made of wood or consist of metal-framed structures with metal siding and corrugated metal tops.

Predator Control

Sheep predators are a real concern, and precautions need to be taken to prevent possible losses. The most common predators are coyotes and dogs and, to a lesser extent, mountain lions, foxes, bobcats, eagles, and owls.

Lambs are the most vulnerable, and extra precautions need to be taken during lambing season, but coyotes, dogs, and mountain lions will kill adult sheep as well. Sometimes dogs will begin to run together in packs, and these packs are known to kill many sheep in one night just for sport.

During lambing season, you can keep your sheep and lambs in a pen and/or barn at night. For pastures, woven wire fencing is recommended to keep the sheep in but is also helpful in keeping predators out. Some predators will try to jump over or dig under the fence. Electrical wires on the outside, along the bottom and top of the fence can be efficient deterrents to prevent the attempts of predators jumping over or digging under.

Another popular way to protect your sheep herd while in the pasture is to use guardian dogs, llamas, or donkeys. The most common dogs used as guardians are Great Pyrenees,

Anatolian Shepherds (also known as the Kangal), and the Maremma Sheepdog. These breeds are well-known for their ability to deter predators.

Two other common means of predator control include shooting or trapping. These can be effective when you have problems with a particular predator that is persistent.

Health Management

Most farm animals need some form of health management as a precaution against possible ailments. The most common diseases for sheep are listed below.

Bacterial Pneumonia

This bacterium is often found in the upper respiratory tract of healthy sheep, causing respiratory infection and death in farm sheep. Symptoms include fever, cough, difficulty breathing, loss of appetite, and lethargy. Treatments include vet-prescribed antibiotics.

Pregnancy Toxemia (Ketosis)

Insufficient nutritional energy during the later stages of the animal's pregnancy triggers the onset of the disease. Symptoms of pregnancy toxemia include reduced appetite, lethargy, delivery, and lambing problems. Treatment includes vet-prescribed propylene glycol.

The likelihood of ketosis can be lessened by ensuring your pregnant sheep have plenty of nutrition in their diet, especially during the later stages of pregnancy.

Coccidiosis

Coccidiosis is a parasite that is typically contracted while grazing or by drinking contaminated water. Symptoms include diarrhea, reduced appetite, dehydration, weakness, weight loss, and the illness can even be fatal.

Treatment includes administering a vet-prescribed coccidiostat, which can be added to feed, mineral, drinking water, and milk replacer. Prevention and control include proper pasture management via rotational grazing and keeping pen areas clean and dry.

Contagious Ecthyma (Orf/Sore Mouth)

Sore mouth is a disease affecting sheep that is transmitted by a virus called *parapoxvirus*. Symptoms include sores and blisters on the face and udder. Treatments include isolation of sick animals and vet-prescribed antibiotics.

Enterotoxemia (Overeating Disease)

This disease is an intestinal infection caused by the absorption of toxins produced by bacteria such as *Clostridium perfringens*. Symptoms include loss of appetite, abdominal discomfort, diarrhea, fever, lethargy, and possible death. Treatments include antitoxins, anti-inflammatories, and oral activated charcoal. Probiotics can be helpful early in the course of the disease. A veterinarian may need to be consulted for proper treatment.

Foot Rot

Foot rot is caused by a bacterium from feces or found in the soil that affects sheep's hooves. Symptoms include limping, grazing on knees, loss of appetite, and possible hoof deformity. Treatment includes isolating infected animals, trimming hoofs of infected animals, administering a footbath solution, and vet-recommended antibiotics. The best preventative measure is to keep the pens and barns clean.

Keratoconjunctivitis (Pinkeye)

Pinkeye in sheep is mainly caused by *Mycoplasma conjunctivae*, targeting the cornea in sheep. Signs of pinkeye include squinting, reddened and swollen eyes, and eye discharge. Treatment involves the immediate separation of sick animals, eye washes, applying Terramycin ointment to the affected eye(s), and administering vet-prescribed antibiotics.

Listeriosis

This disease is caused by ingesting spoiled or moldy forages. Symptoms in infected sheep include staggering, partial nerve paralysis, ears drooping, salivation, problems with swallowing, and the illness can cause possible death. Treatment includes vet-prescribed antibiotics. Prevention can include discarding moldy or spoiled feed and hay.

Expenses

Purchase Cost

The average purchase cost for a sheep, ranges from $200 to $300 *according to the* American Sheep Industry Association. The price largely depends on the breed, age, and weight.

Breeding Costs

The cost of breeding will be dependent on whether you choose to own your own ram or if you choose to pay for the breeding service each year. The price for a ram will be, on average, about $350–$500. A single ram can service up to 40 ewes. Having your own rams is generally the best option for those with medium-to-large herds.

For those with small herds, the possibility of using borrowed rams as a breeding service and the use of artificial insemination are both viable options if they are available in your area. The prices for either of these options vary depending on animal numbers.

Feed Cost

Feed for sheep will generally consist of both hay and grain when not grazing on pasture. These prices can vary widely depending on the area and time of year.

Hay Cost per Bale

It is OK to feed sheep a mixture of grass hay and legume hay during gestation but is best to feed mostly alfalfa hay during lactation. Depending on the area, alfalfa hay costs $200–$245 per ton, and grass hay about $140–$240 per ton. That would translate to be $9–$11 per 90-lb. bale for alfalfa, with an average of about $10 per bale. Grass hay is roughly $7 per bale. Again, these prices can vary widely per geographic area and time of year.

Grain Cost per 50-Lb. Bag

Commercial livestock grain usually comes in a 50-lb. bag. The best grain feed for sheep contains a mixture of ground corn, wheat, barley, rye, oats, soybean, and cottonseed. The average cost for a 50-lb. bag generally runs about $22.

Daily Feed Cost

An average 140-lb. ewe will eat about 4–5 lb. of forage per day. On average, grass hay costs $7 per 50-lb. bale, and alfalfa costs $10 per 90-lb. bale. If you were to mix the two, the overall average cost to feed 1 ewe 5 lb. of forage would be roughly $0.65 per ewe per day. The weight per bale and price per bale can vary widely depending on the area, time of year, and other economic factors.

In addition, a lactating sheep should also have grain as part of the diet. The amount of grain should be formulated by using 1.5% of body weight. For a 140-lb. ewe, that would be 2 lb. of grain per day.

- 50-lb. grain bag @ $22 = $.44 per lb.
- $.44 x 2 lb. per day = $.88 per day

Hay cost =	$.65 per day
Grain cost =	$.88 per day
Total cost =	$1.53 per day

Feeding Equipment

Feeding equipment such as feed troughs and feed bunks can be purchased or fabricated on-site. Such items can be fabricated out of a variety of materials, but for long-term usage, troughs and feed bunks are typically fabricated using metal or wood. Metal or plastic versions can be purchased online.

Troughs

On average, a feed trough will cost roughly $20–$40 per linear foot. The typical size is 12 in. wide, 8 in.–9 in. deep, and 6 ft. long.

Feed Bunks

Feed bunks will run between $60 and $100 per linear foot. The typical size is about 4 ft. wide, 5 ft. long, and 5 ft. high.

Feed Storage

Feed storage can be in a shed, barn, or shipping container. A good 8 ft. x 40 ft. shipping container can be purchased for $2000.

Bedding Cost

Bedding should be provided in the shed or barn during the winter and lambing season. The lambing pen especially should be a clean and dry space with bedding. This can be done with straw bales. One straw bale costs approximately $5 per bale, and one bale should last a week.

Watering Equipment

Water containers can come in a variety of shapes, sizes, and materials. The most common material is either plastic or galvanized metal. The common shapes are either round or oblong.

Plastic Water Troughs

A 40-gallon plastic oblong-shaped watering trough will cost $60 ($1–$1.5 per gallon). Larger troughs will run closer to $2 per gallon.

Galvanized Water Troughs and Tanks

An oblong water trough, with the dimensions of 2 ft. x 2 ft. x 4 ft. (44 gallons), will run about $110 ($2.5 per gallon). A round galvanized water tank of 2 ft. x 3 ft. (53 gallons) costs $120 ($2.26 per gallon).

Containment Costs

The cost of fencing is another mandatory expense. On average, corrals or pens around the heart of the homestead where the sheep are kept and fed will generally consist of at least woven wire fencing with metal posts. Some homesteads like to make their corrals/pens of wood or welded pipe for better structural integrity, and then add woven wire fencing to make the containment sheep-proof. Pasture fencing is normally constructed with woven wire and metal posts.

Wooden Fence Cost per Foot

Split rail wooden fences will cost $8–$12 per linear foot. Labor will cost ~$10 per linear foot.

Pipe Fencing Cost per Foot

Pipe fencing will cost $10–$15 per linear foot. This is more costly than a wooden corral but will last much longer as well.

Woven Wire Fence Cost per Foot

Regardless of whether wood or pipe is used, woven wire fencing will also be needed. The average cost for a woven wire fence with metal posts will depend on the type of mesh, but on average will run about $1.50–$1.90 per linear foot. This type of fencing is commonly used for pasture fencing.

Electric Fencing

Electric fencing for livestock generally runs about $1 per linear foot. This type of fencing can be a good choice for pasture divisions when sheep are moved on a monthly or bimonthly basis.

Shelter

You should have a shed or barn for shelter and for lambing pens. Simple sheds can cost $15 per sq. ft. Metal barns can cost $25–$50 per square foot. Wooden barns, on average, cost $45–$65 per square foot to build.

Vet Care

It is not uncommon to require a veterinarian's service occasionally. You might need treatment for a disease or to assist with AI. Therefore, it may be wise to include this as a possible expense item in your annual budget. Medicine and vet care can be $5–$10 per month per sheep.

Start-Up Expense Summary Charts

The following expense summaries are designed to provide a simple example for demonstrating start-up expenses. The following charts will be based on initial purchase of a 140-lb. sheep for a homestead with 1 acre of pasture. Fencing and feeding equipment are included. The livestock pen will be 60 ft. x 40 ft., with two 20 ft. x 20 ft. internal pens. An enclosed shed (800 sq. ft.) will be used for shelter. The 1-acre pasture will be enclosed with woven wire fencing.

Initial Livestock Investment Expense Chart

Item	Cost
Per 140-lb. ewe	$250

Infrastructure and Equipment Expense Chart

Item	Cost
Woven wire fence 60 ft. x 40 ft.	200 @ $1.75/ft. = $350
Two interior pens 20 ft. x 20 ft.	60 ft. @ $1.75/ft. = $105
One-acre perimeter fenced with woven wire fence & metal posts	836 @ $1.75/ft. = $1463
Shipping container for feed	$2000
Barn (800 sq. ft.)	@ $25/sq. ft. = $20,000
Feed trough (6 ft.)	$180
Feed bunk (5 ft.)	$400
Water trough (40 gallons)	$60
Electric clippers	$250
Total	**$24,808**

Operating Expenses Summary Chart

The operating expenses are based on maintenance per 140-lb. ewe, for a homestead with 1 acre of pasture and grazing for 7 months. Hay will be fed during the winter and spring, and grain will be fed during the last month or two of gestation and during lactation.

Feed Costs

Item	Daily Cost	Monthly Cost	Annual Cost (5 Mo. Feed)
Grass & alfalfa hay (5 months)	$.65	$19.50	$97.50
Grain (5 months)	$.88	$26.40	$132.00
Minerals (5 months)	$.10	$3.00	$15.00
Total	**$1.63**	**$48.90**	**$244.50**

Animal Care

Item	Annual Cost
Bedding (4 mo. winter & lambing)	$80
Breeding cost	$50
Vet care	$60
Total	**$190**

Revenue

Sheep, being a dual-purpose animal, can provide revenue from both wool and lambs. Ewes are generally shorn once per year. The average weight per fleece is about 12 lb. The average price for raw wool for 2022–2023 was $1.70. At $1.70 per pound, the revenue for a 12-lb. fleece is $20.40.

Market lambs are usually sold when they reach about 90 lb. Ewes typically have 2 lambs. The average price per pound for market lambs is $250 per cwt. That is an average of $225 per lamb. This will yield $450 for 2 market lambs per ewe (90 lb. x $2.50 x 2).

The revenue for meat would be roughly the same. The dressing percentage is 50%, so each 90 lb. lamb would yield a 45 lb. carcass. At $5 per lb. for lamb meat, the yield for 2 carcasses would be 90 lb. meat x $5 = $450.

The overall annual revenue of wool plus lambs is roughly $470.40 per ewe.

Profit / Loss Analysis

Profit/Loss

The following chart compares the expense of feed and animal care versus the revenue from wool and lambs.

Please note that sheep purchase cost and infrastructure cost are not included. The same infrastructure can be used for other livestock options, and more than one year.

Item	Annual Expense (1 Ewe)	Annual Revenue (1 Ewe & Lambs)	Annual Profit/Loss
Feed & misc.	$244.50		
Animal care	$190.00		
Total expense	**$434.50**		
Wool		$20.40	
Lambs		$450.00	
Total revenue		**$470.40**	
Profit/loss			**$35.90**

The best way to gain more profit is to implement one or more of the following.

- Reduce feed costs with more grazing if possible.
- Increase wool profit with wool processing and marketing.

Profitability

Sheep are one of the more prevalent livestock animals in the US when it comes to total numbers. In reference to the sheep industry as a whole in the US, sheep are primarily raised on farms and ranches that have large acreages for grazing in the western states. The main reason for sheep's popularity is that they have a dual purpose allowing them to be raised for both meat and wool.

The price of wool is much lower than mohair or cashmere, and sheep are shorn only once per year. Sheep often have twin lambs, which helps with the overall expenses, but the key to profit is grazing 8 or more months per year, producing your own hay, and buying grain feed in bulk.

If your space is limited, and you have to pay for feed (hay and grain) at local prices, about the most you can hope for is to break even with wool and lambs. If you have limited space and a small herd, you can try to have your wool processed in order to generate more revenue to offset expenses.

Picture Gallery

Delaine-Merino

Rambouillet

Debouillet

Southdown

Hampshire

Suffolk

Columbia

CHAPTER 5

Pigs

I f you want to raise livestock for meat, with the potential to generate revenue, then raising pigs is definitely worth considering. The reason pigs are a viable option for meat and revenue is because they can have two litters per year and can average 10 piglets per litter. Another reason they are such a unique farm animal is because they are omnivorous and can eat just about anything.

Under the formal category of swine, these farm animals are most often referred to as either pigs or hogs. In this chapter, the most common breeds used for homestead hog rearing are covered as well as breeding, feeding, animal care, farrowing, containment, shelter, and health management. This chapter will conclude with expenses, revenue potential, and a profit/ loss analysis to assist you in the decision-making process.

Breeds

The following breeds are a favorite on small farms and homesteads. These breeds are popular for their temperament, litter size, mothering ability, meat quality, growth rate, and foraging ability.

Berkshire

Berkshire pigs are black with white markings on their legs, face, and tail. The meat is a favorite with meat markets, due to its great marbling and flavor. Because of their calm temperament, they are often considered suitable for both contained systems and pasture-based systems.

Chester White

Chester White pigs are known for their white color, stout build, and floppy ears. They typically produce large litters, and being good mothers, they excel at raising piglets.

Duroc

Duroc pigs are reddish-brown in color. They are commonly used in commercial pork production because of an efficient feed-to-meat conversion and fast growth rate. The Duroc also has a good meat-to-fat ratio, causing their meat to be flavorful and tender.

Hampshire

Hampshire pigs are black with a white belt around their shoulders and front legs. They have a stocky build with erect ears. They are known for their good meat quality and efficiency in feed conversion. They adapt well to various management systems.

Gloucestershire

These pigs have a distinctive appearance due to their white body with black spots. They are known for their calmness and friendly nature. Gloucestershires are valued for their flavorful meat. Because of their foraging abilities, they tend to do well in pasture-based systems.

Yorkshire

Yorkshires are large, white pigs, having a straight build and erect ears. Yorkshires are often used in commercial pork production because of their fast growth and efficient feed conversion. Their meat is a leaner cut and considered high in quality. They generally have a larger litter than other breeds (10–14 piglets in contrast to the typical 8–12 piglets).

Breed Selection Considerations

When selecting pig breeds for a small farm, several criteria should be considered to ensure that the chosen breed aligns with your goals, resources, and management capabilities. Here are some of the most common breed selection criteria for small farm pigs.

Size

Size is often an important factor to consider if you have a small farm. Smaller breeds can be more manageable when space is limited. For farms with more space available, size isn't as much of a factor.

Temperament

If you are a beginner or if you have family members interacting with the pigs, breeds that are more docile and have calm temperaments are often better choices for small farms.

Adaptability

A breed that can adapt to a variation of hot and cold temperatures is a great plus. Also, some breeds are better suited to outdoor systems and foraging, while others can do well in confinement. It is a great advantage to choose a breed that can adapt to either type of operation.

Litter Size

Larger litters can lead to greater profitability, so it is important to consider litter size as an important trait when you choose a breed for your operation.

Mothering Abilities

Good mothering ability is a trait that is crucial for piglet survival and growth. A good mother will not only provide adequate milk but will also try to keep her piglets warm and protect them from harm.

Ease of Farrowing

Breeds that do well during farrowing will often have litters with higher survival rates and will less likely require veterinary assistance during delivery.

Meat Quality

If you are raising pigs for meat, factors like marbling, tenderness, and flavor are considered important criteria. Some breeds excel in producing high-quality pork.

Growth Rate

Growth rate and feed efficiency are important metrics of fast-growing pigs for meat production—especially if a shorter time to market is your goal.

Foraging Ability

If you plan to use a pasture-based or free-range system, you should look for a breed that will have strong foraging instincts and can efficiently convert natural vegetation into meat.

Health and Hardiness

Choose breeds with a reputation for good health and disease resistance. Pigs that require fewer veterinary interventions can save you time and money

Comparison Chart

All the breeds performed well with each criterion, and all were similar in performance. There were some, however, that excelled a bit above the others in certain areas. Those criteria and high performers are listed below.

Consideration Criteria	High Performers
Adaptability	Hampshire
Litter size	Yorkshire
Mothering ability	Chester White
Meat quality	Berkshire
Growth rate	Duroc
Foraging ability	Gloucestershire

Gestation

The gestation period for a sow is approximately 114 days, which is close to 4 months. This period refers to the time between successful mating and giving birth to a litter of piglets. Keep in mind that variations can occur, so it's important to monitor the sow's condition and be prepared for farrowing when the time comes.

Breeding

With a gestation period of 114 to 116 days, it is common to have two litters per year, with adequate rest between litters. Sows will come into heat every 21 days, which will last 2–3 days.

If you have more than 1 sow, it is often feasible to have your own boar for servicing, because the sows can have two litters per year.

In most cases, it's recommended to keep boars and sows separated except during the specific mating period. There are several reasons for this separation.

1. Minimizing injuries: Due to the boar's larger size and weight, it is possible that the boar could harm the sow

while attempting to mount her during non-heat periods. Separating them also helps prevent fight-related injuries.

2. Controlled breeding: Keeping the boar and sow in separate pens helps to control the timing and thus ensure that mating occurs at the optimum time of her heat cycle.

3. Feed management: It is also easier to provide the best type of feed and ration for both the boar and the sow by keeping them in separate pens.

Generally speaking, keeping boars and sows separated when not in heat will lead to better reproductive success and overall better health.

Breeding and Lactation Cycle Timing

The gestation period for the sow is 114–116 days (roughly 4 months). The litter will suckle the sow for 2–3 months. After the piglets are weaned, the sow will normally come into heat within a few days and can be bred again at this time. With this schedule, you can arrange to have two litters each year, farrowing in the spring and the fall.

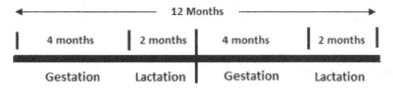

Feeding

Using the right types of feed and amounts is crucial for meeting nutritional needs and maintaining healthy growth and development. As your litters get older and gain weight, their needs for protein and energy will change.

You can use commercial feed, which is commonly divided in the following groups, or you can use alternative feed purchased in bulk. Other supplemental food sources from your homestead can also be used, such as damaged or surplus fruits and vegetables, table scraps, dairy by-products, etc.

Commercial Feed

A common way to ensure that the hog's nutritional needs are met is to adjust the protein level and grain rations of the feed at certain hog age and weight levels. The protein needs will slowly decrease with age and weight. Protein helps support muscle development and overall growth.

The energy content of the feed remains important, and the rations will change as the pigs get heavier. Grains like corn, barley, and wheat are commonly used to provide energy and for weight gain. Fats and oils can also be included. The feed should contain adequate vitamins, minerals, and amino acids to support bone health, immune function, and overall health.

Pelleted or granulated feeds are often recommended as they are easy for young pigs to consume, and they minimize waste.

Gestating Sow

Feeding a proper diet and ration are important for the sow's health and the health of the developing fetuses. Feed for gestating sows should have a protein content of about 12–14%. The diet should be formulated to meet energy needs, but not lead to excessive weight gain. Grains and corn are common components for energy.

Fiber in the diet can help them feel satisfied and thus reduce the risk of gaining too much weight. Gestating sows may require specific mineral supplementation, such as calcium and phosphorus, in order to address the needs of developing fetuses.

Pelleted or ground feeds are commonly used for gestating sows. Typically, gestating sows are fed once or twice a day. On average, they might consume about 4–6 lb. of feed per day. The sow should receive 4 lb. per day for the first 74 days, and 6 lb. per day during the last 40 days (during major fetus development). Adequate water intake supports overall health and reproductive success.

Lactating Sow

In order to support milk production and maintain body condition, lactating sows have high nutrient requirements. Their diet should contain about 14–16% protein or more. Adequate protein is vital for milk production and maintaining body weight. The diet should contain corn and other energy sources to achieve the needed energy requirements.

The diet should contain a balance of essential vitamins, minerals, and amino acids. These nutrients are essential for piglet health, and, in order to maximize digestibility, pelleted or ground feeds are often used.

Lactating sows should be fed multiple times a day—typically in two to three feedings—or have access to an automatic feeder. On average, the sow might consume 14–16 lb. of feed per day during lactation. For the first 2 weeks, the sow should receive roughly 4 lb. of feed per day plus 1 lb. per piglet. For the remaining weeks until weaning, the sow should receive 6 lb. of feed per day plus 1 lb. per piglet.

Accessibility to clean, fresh water at all times is important to lactating sows. Water intake is closely linked to milk production and overall well-being.

Creep Feed (14–40 Lb.)

Creep feed is made available to the piglets at an age of 10–14 days and is fed until weaned. The feed is high in protein (21–23%) and is formulated to help the piglets transition from milk to solid food. The pellets are extra small to accommodate the preweaning age. During this time frame, the piglets will consume, on average, about .25–.8 lb. of feed per day.

Starter Group (40–60 Lb.)

Pigs at this stage require a feed with a high protein content. Starter feeds typically contain about 18–20% protein. High-quality protein sources, like soybean meal, are commonly used.

Pigs at this weight are usually fed three to four times a day or have access to an automatic feeder. The amount of feed is about 2–3% of body weight per day. For pigs weighing 40–60 lb., this would translate to roughly 0.8–1.5 lb. of feed per day.

Starter-Grower Group (60–100 Lb.)

As pigs grow, their requirements for protein, energy, and overall nutrients will change. The protein content in the feed should be about 16–18% at this stage. The energy content of the feed (grains) remains important, and rations for such will increase.

Pigs in this weight range can be fed three to four times a day or have access to an automatic feeder. The recommended feeding rate can range 2–3% of their body weight per day. For pigs weighing 60–100 lb., the ration would be approximately 1–2.4 lb. of feed per day.

Grower Group (100–160 Lb.)

For the grower diet, the protein content should be 14–16%. These slightly lower protein levels can be used as the pigs continue to grow and their muscle development slows a bit. This diet should continue to provide energy-rich portions of grains to support growth and to add weight.

Pigs in this weight range can be fed two to three times a day or have access to an automatic feeder. The recommended feeding rate can vary but is generally about 2–2.5% of their body weight per day. For pigs weighing 100–160 lb., the ration would be approximately 2–4 lb. of feed per day.

Finisher Group (160–250 Lb.)

The finisher feed is designed to achieve optimal growth as the pigs near market weight. The protein content in the finisher feed is typically about 12–14%. These slightly lower protein levels can be used because the pigs' growth rate has generally slowed. The energy-rich diet—using grains like corn, barley, and wheat—is still important for achieving desired weight gain.

Pigs in this weight range can be fed two to three times a day or have access to an automatic feeder. The recommended feeding rate can vary but is generally about 2.5–3.5% of their body weight per day. For pigs weighing 160–250 lb., the ration would be approximately 4–9 lb. of feed per day.

Feed Alternatives

Commercial feeds can be purchased for the various weight groups listed above at any local feed store. The feed typically comes in 50-lb. bags and is advertised for starters, growers, and finishers. However, feed can also be purchased as complete pig feed, designed for a wider range of weights.

Because commercial feed is so expensive, it is common to look for alternatives. One alternative to commercial bagged feed is to purchase grain in bulk from a mill, which generally comes in 100-lb. bags. The overall price per pound can be half the price of commercial bagged feed.

You might be able to find these available already mixed, or you might have to purchase the various grains (wheat, barley, corn, etc.) separately and either arrange for them to be mixed or mix them yourself.

If you want them granulated, you might be able to find bulk grain in this form or arrange for the grain to be ground for you or grind it yourself. Of course, you would need to have access to the appropriate equipment.

For protein, you might be able to find a source for soybean meal or whole soybeans that will need to be ground into meal for you or grind it yourself. Alfalfa can be purchased in cube form much cheaper than bagged pellet form. These forms of protein can then be mixed with the bulk grains. You can find the appropriate formulas online.

Supplemental Food Alternatives

Hogs are omnivorous and will eat just about anything. If you are growing your own nut and fruit trees and/or have a garden, much that is often discarded can be used for supplemental pig feed. It is common for homesteads to supplement their hog feed with available food items, such as surplus or damaged fruits and vegetables, dairy by-products (like buttermilk and whey), table scraps, lawn clippings, wasted hay from goat and sheep feed bunks, etc.

Some homesteads try to offset feed costs by allowing pigs to forage in a fenced pasture. Pigs can forage on grass, but they have a single stomach and, therefore, aren't as efficient at getting nutrition from grazing as sheep and cows. They can also cause a lot of damage, as they dig up the ground looking for roots.

If you have extra land for foraging, a better use of the land for hog feed would be to raise a garden. Hogs can eat all parts of the vegetables, including the green tops, and this is an excellent way to meet their nutritional needs for vitamins.

Feeding Equipment

Feeding equipment needs to be heavy to prevent the pigs from overturning the trough or feeder. For this reason, the feeders or troughs are generally made of heavy gauge metal. The feeder or trough can also be anchored with metal stakes to prevent it from being overturned.

You can have open feed troughs, which you fill by pouring the feed rations from a bucket, or you can use an automatic feeder. The automatic feeders generally consist of a covered upright feed bin (either rectangular or round) and a feed trough underneath.

The upright feed bin is filled with feed (pellet or granulated). The feed in the bin fills the trough underneath. The trough for the rectangular version is generally covered with a metal flap to protect the feed from being spoiled by rain and to keep rodents out.

For the covered trough versions, the pig will flip the flap up with their nose and eat from the trough until they are satisfied. As the feed is consumed from the trough, feed is replenished from the upright bin.

Watering

Daily water requirements will depend largely on the hog's age and weight. For instance, a starter (40–60 lb.) will drink 1–2 gallons per day; a grower (60–100 lb.) will drink 2–3 gallons per day; a finisher (100–250 lb.) will drink 3–5 gallons per day; and a lactating sow will drink 4–7 gallons per day.

Water intake can fluctuate based on temperature, humidity, feed type, activity level, and other factors. During hot weather or lactation, water intake can be higher due to increased water needs for cooling or milk production.

Adequate water is crucial for hogs for their digestion and to regulate body temperature. In hot weather, they drink more water to cool down and prevent heat stress.

For lactating sows, adequate water intake is essential for milk production to nourish piglets. For breeding sows and boars, proper water intake supports reproductive performance and fertility.

Water intake is also closely linked to feed consumption. Hogs that have access to water tend to consume more feed, leading to greater weight gain and growth. In essence, sufficient water contributes to efficient growth, optimal feed conversion, and better production outcomes.

Watering Equipment

It is possible to use almost anything that would hold water, but it isn't recommended. Water troughs that are used for other types of livestock don't really work well for hogs, because the water will soon become dirty.

For that reason, most pig farmers use automatic watering equipment. These types of water systems allow the pig to get cool, clean water when they need it, without having an open water container. There are two common types of automatic watering systems that are most often used: (1) the automatic water nipple and (2) the automatic water cup.

The water nipple is a round pipe-shaped fitting, which has a solid metal rod in the center. When the rod is moved up or down, water is allowed to flow. The pig will put their mouth over the nipple and cause the rod to move, thus allowing water to flow into their mouths. When released, the water stops flowing.

The water cup is an oblong-shaped metal cup, with a metal flap at the back of the cup serving as a float valve. When the hog pushes the flap with their nose, water flows into the cup, allowing the pig to drink from the cup.

Farrowing

Farrowing is when a sow gives birth to a litter of piglets. Much has been written about farrowing and how to manage and prepare for that important and exciting event. In this section, the various types of equipment used to manage the farrowing event will be discussed.

There is a lot of controversy about the types of equipment that should be used during farrowing—mostly regarding the use of gestation crates and farrowing crates. The types of farrowing methods and equipment that are more commonly used in homesteading are reviewed in this section.

There are many ways to arrange for pig farrowing, and there are many ways to construct farrowing equipment. The following methods and construction models are only one of many ways of accomplishing the goal.

Farrowing Pens

The use of farrowing pens is more common on small farms and homesteads. The farrowing pen is a specialized enclosure designed to provide a controlled and comfortable environment for the sow to give birth (farrow) and nurse her piglets until they are weaned. The pen is arranged for the piglets to have protection and warmth but also allows adequate space for the sow to turn around, maneuver, and interact with the litter. The sow and her litter generally use this pen until the litter is 4–8 weeks old.

Wall Construction

The farrowing pen is a square enclosure that is roughly 8 ft. x 8 ft. or 64 square ft. The walls can be made of pipe, with hog panels attached to the inside to prevent the piglets from escaping, or the walls can be partially solid (24" high with 2' x 8' plywood and pipe above), or the walls can be entirely solid (i.e., with a 4' x 8' sheet of plywood).

Rails

Horizontal rails need to be added to the walls to prevent the sow from accidentally crushing her piglets while lying down. The horizontal rails need to be about 20 in. off the floor and extend outward about 12 in.

Water and Feed

Easily accessible feeding and watering systems need to be installed in one corner of the pen for the sow to ensure proper nutrition during lactation.

Creep Area

A special space for the piglets—called a creep—needs to be built to protect them from being stepped on or crushed when the sow is lying down and to also keep the piglets warm when not suckling. The area is generally constructed in the corner adjacent to the sow's feeder.

The area is normally created by constructing a short wall in the corner to create a triangle-shaped space with exit and entrance gaps. The space would be roughly 30" x 30" x 42". A heat lamp is installed above the creep area to keep the piglets warm.

After the piglets are 1 week old, they can be started on feed and water, which is also made available in the creep area. This is called creep feeding.

Flooring

The flooring can be arranged in a variety of ways. A common method is to use concrete as the flooring material, so that the area can be kept clean more easily. Another common arrangement on small farms is to have dirt floors but this requires careful management and the use of fodder to keep the floor sanitary and safe.

The use of crushed granite on top of dirt is one way to manage the floor so that it will not become muddy or slippery. This type of flooring can remain flat and hard yet allows for proper drainage to manage waste and maintain a sanitary environment. If crushed granite is used, an additional layer of fodder, like sawdust, should be used on top of the crushed granite for added comfort.

Another often-used alternative flooring is rubber mats. There are rubber mats that are especially designed for farrowing pens, which allow for good traction, comfort, and good drainage.

Regardless of the type of floor material you choose to use, the floor should be kept as clean as possible, by cleaning out feces and soiled fodder as often as necessary and replenishing fresh fodder if used.

Nesting and Floor Fodder

Just before giving birth, the sow wants to provide a safe, secure space for herself. This behavior is called nesting. The most common method is to provide a bale of clean, fresh grass hay on the floor the day of parturition. The sow will arrange the fodder as she wants for the nest.

After parturition, it is a common practice to take the grass hay out and replace it with wood shavings. This allows the little piglets to navigate more easily when wanting to suckle or get out of the sow's way when she is moving or lying down.

Space Management

Typically, farrowing is done indoors, and oftentimes an outdoor pen is arranged so that the sow and piglets have the opportunity to go outside during the day in good weather, after the piglets reach an age of 10–14 days. Later, after the piglets are weaned, the sow and piglets are placed in a separate pen outdoors.

Most often, the boar is kept in a separate pen for breeding timing purposes. The sows are often kept in a separate pen from the piglets after weaning, because they are generally fed different types of feed with differing protein percentages.

Containment

Fencing needs to be strong to withstand the strength and weight of adult hogs. It also needs to incorporate the right kind of material in order to keep small piglets from escaping.

Panels and Posts

Hog panels are used more than any other form of fencing material. A hog panel is made of heavy gauge galvanized ¼ in. metal rods. Panel dimensions can come in 16 ft. or 8 ft. lengths and are generally either 34 in. or 50 in. tall.

The rods are welded to form a mesh pattern, with the mesh openings smaller at the bottom and larger at the top. This arrangement prevents the piglets from escaping through the mesh openings.

Hog panels are supported and attached to posts. The most common version is to use metal T posts, spaced every 4 ft. Corner posts should be 8" x 8" x 8' wooden posts, to provide more strength.

Separate Pens

It is quite common to have a separate pen for the boar, for the sow after weaning, and for feeder pigs (the weaned litter). The minimum pen size for the boar and the pen for the sow are roughly 8 ft. x 16 ft. (roughly 128 sq. ft.). These pens can be larger, if you have more space. The minimum pen size for the weaned litter of 10 pigs can be roughly 16 ft. x 40 ft. (64 sq. ft. for each pig), but a larger pen should be provided if you plan on finishing the pigs beyond 100 lb.

All three pens should provide shelter at one end of the pen to protect the pigs from heat and harsh weather.

Shelter

Hog shelters, also known as pig housing, serve the purpose of providing protection, comfort, and a controlled environment for pigs raised in various agricultural settings. Shelters can be made of metal or wood, but metal sheds or barns are the most common.

These shelters are designed to meet the specific needs of pigs, ensuring their well-being, health, and optimal growth. They offer the following benefits.

Protection from Weather

Hog shelters shield pigs from adverse weather conditions, such as extreme heat, cold, rain, and wind. This protection helps prevent stress, illness, and other health issues associated with exposure to harsh elements.

Temperature Regulation

Temperature control is an important factor in raising pigs. Shelters should be insulated and equipped with ventilation systems to help maintain a comfortable temperature range for the pigs. This is particularly important for young piglets and sows during farrowing. Heat control will also help to ensure regular feed intake.

Comfort and Behavior

Pigs are more likely to exhibit natural behaviors and better growth rates when they are protected from heat in a sheltered environment. Adequate space, flooring, and bedding also contribute to their day-to-day health.

Efficient Feeding and Management

Outside pens usually have feeders and waterers available, but feeding and water supply systems can also be made available inside the shelter when the need for protection from harsh weather can be long-lasting. Pigs won't eat if they are too hot. If they aren't eating, they aren't gaining weight, which you want for your feeder and grower pigs.

Optimal Growth

With proper protection and care provided by shelters, pigs can achieve their growth potential more effectively, leading to higher-quality meat production.

Ease of Handling

Shelters are often designed to facilitate easy handling and management of pigs for tasks such as health checks, vaccinations, and moving animals between different stages of production.

Health Management

It is important to have a health management plan to keep your hog operation healthy and safe. It is much easier and economical to vaccinate pigs for possible diseases, than to have to treat them after they have contracted a disease.

Vaccination

Here are seven commonly used vaccines that could be considered for a small farm hog operation.

1. Porcine Circovirus Type 2 (PCV2) vaccine: This vaccine is to combat respiratory and reproductive issues.
2. Parvovirus vaccine: *Parvovirus* can cause reproductive issues in sows, including stillbirths and mummified fetuses.
3. Mycoplasma vaccine: *Mycoplasma* is a common respiratory pathogen in pigs that can lead to reduced growth and respiratory issues.
4. Erysipelas vaccine: *Erysipelas* is a bacterial disease that can affect pigs, causing sudden death, skin lesions, and lameness.
5. Swine Influenza vaccine: *Swine influenza* can cause respiratory illness in pigs, affecting their growth and overall health. Vaccination against swine flu can be important, especially if your pigs are at risk of exposure.

6. *Clostridium perfringens* Type C vaccine: This vaccine is for enteric diseases and diarrhea.

7. Atrophic Rhinitis vaccine: This vaccine targets bacteria that can contribute to *atrophic rhinitis*, a condition that affects the nasal passages of pigs.

Note: It is really important and highly recommended to consult with a veterinarian and create a health management plan and vaccination program that will address the most relevant health concerns for your small farm hog operation.

Deworming

Hogs, like any other livestock, should be dewormed to maintain their health and prevent the negative effects of internal parasites. The timing and frequency of deworming can depend on several factors, including age, environment, and the presence of parasites in the area.

The most common internal parasites in hogs include round-worms, whipworms, and coccidia. Symptoms of a parasite problem can include poor growth, diarrhea, and lethargy.

Piglets are often more susceptible to internal parasites, so a particular deworming program should be initiated for them. Piglets can be dewormed at about 2–3 weeks of age, and then every 4–6 weeks until they reach a certain age (typically 6 months) when they have developed more resistance.

If your pigs are kept in outdoor or pasture systems, they might have more exposure to parasites due to the environment. Parasites in your pigs can be determined by way of periodic fecal testing.

It is often recommended to rotate between different classes of deworming treatments from season to season, in order to prevent the possibility of parasite resistance to specific deworming medications.

Note: It's important to work closely with a veterinarian when developing a deworming program for your pigs.

Expenses

Breeding Stock

The purchase cost for breeding stock for hogs in the US can vary quite a bit. The price range listed below is for average prices and largely depends on the breed, age, weight, genetics, time of year, and locale of purchase.

Your entire hog operation will depend on the quality and efficiency of your breeding stock. With that in mind, it is a good idea to purchase breeding stock from reputable breeders who have a proven track record and documented lineage of quality stock.

This can be a substantial initial expense, but starting off with quality stock with good genetics will be worth the investment. Good breeding stock can run from $350 to $2000 or more, but the average price range is $350–$750 per pig.

Feed Cost

Feed cost is variable depending on the types of food fed (i.e., commercial feed [retail], mill feed [wholesale], supplemental food, etc.). In order to find the total feed cost per pig, there are several things that need to be calculated from accurate records (i.e., feed for the sow during gestation and lactation periods, the size of the litter, and the amount of feed for each pig during the various stages of growth).

Feed for the sow is based on certain rations for a certain amount of time. The amount of feed for the litter is determined from daily rations based on the age, weight, and stage of growth. The following figures are based on commercial retail prices averaging $25 for a 50-lb. bag of feed. All of the stages of protein rations cost roughly the same per bag.

Wholesale feed can be purchased in 100-lb. bags from a feed mill, for roughly half the cost of commercial feed. If you have access to such a feed mill, you can use figures from the following charts and divide them by 2 to get an idea of cost.

Obviously, the amount of retail or wholesale feed you use can be reduced by the use of homestead supplemental feed, but that is difficult to measure, so these figures will be based on purchased feed only.

Boar Feed

The amount of feed that a full-grown boar hog eats per day can vary and depends on several factors, including the hog's size, age, and breed. On average, a mature boar hog may consume between 4 and 8 lb. per day, but the average should fall about 6 lb. of feed per day.

For this example, the cost of the boar's feed is not included when computing the cost of feed per pig per litter in the charts below—because you may choose to use artificial insemination.

Sow Feed during Gestation

The gestation period is roughly 4 months. For the first 74 days, the sow should be fed 4 lb. per day, increasing to 6 lb. per day for the last 40 days. For the duration of the 4-month gestation period, the sow should be fed roughly 536 lb. of feed. At $0.50 per lb. for a 50-lb. bag of commercial retail feed, the total will be $268.

Animal/ Stage	Average Age Range	Days Fed	Average Daily Ration	Total Lb. Fed	Retail Feed Cost ($.50/Lb.)
Sow— gestation	N/A	74 days 40 days	4 lb./day 6 lb./day	296 lb. 240 lb. = 536 lb.	**$268**

Sow Feed during Lactation

In this example, the lactation period is 8 weeks with a litter of 10 piglets. During the first 14 days, the sow will be fed 14 lb. of feed per day (4 lb. plus 1 lb. for each piglet). For the remaining 42 days, the sow will be fed 16 lb. of feed per day (6 lb. plus 1 lb. per piglet).

Animal/ Stage	Average Age Range	Days Fed	Average Daily Ration	Total Lb. Fed	Retail Feed Cost ($.50/Lb.)
Sow— lactation (2 weeks)	N/A	14 days	4 lb./day (+ 1 lb./ piglet = 14 lb.	196 lb.	$98
Sow— lactation (6 weeks)	N/A	42 days	6 lb./day (+ 1 lb./ piglet) = 16 lb.	672 lb.	$336
Total				868 lb.	$434

Pig Litter Feed

There were 10 piglets in the example litter, so the feed cost for each piglet begins by assuming its litter share (one-tenth) of the sow's total feed cost. Thereafter, the feed cost for each piglet is calculated and accumulated for each age/weight stage. The daily rations per piglet are based on averages. The running total shows the total feed cost per piglet, at the end of each stage. This will allow you to see what you have invested if you choose to sell the pigs at the end of a certain age/weight stage.

Animal/ Stage	Average Age Range	Days Fed	Average Daily Ration	Total Lb. Fed	Retail Feed Cost (.50/Lb.)	Running $ Total per Pig
Total sow feed cost per piglet						$702 ÷ 10 = **$70**
Pig–creep 14–40 lb.	.5–2 months	42 days	.5 lb./day	21 lb.	$10.50	**$80.50**
Pig–starter 40–60 lb.	2–3 months	28 days	1 lb./day (.8–1.5 lb.)	28 lb.	$14.00	**$94.50**
Pig–starter-grower 60–100 lb.	3–5 months	56 days	1.7 lb./day (1–2.4 lb.)	95 lb.	$47.50	**$142.00**
Pig–grower 100–160 lb.	5–7 months	70 days	3 lb./day	210 lb.	$105.00	**$247.00**
Pig–finisher 160–250 lb.	7–10 months	70 days	6.4 lb./day (4–8.75)	448 lb.	$224.00	**$471.00**

Rule of Thumb for Feed Cost

The rule of thumb to determine the cost of feed necessary to achieve any weight is to use the following formula: Figure 3 lb. of feed necessary for every 1 lb. of gain, then multiply that amount by the cost of feed per pound. For instance, if you were to sell a starter pig at 100 lb., the results would be thus: 100 x 3 = 300 lb. of feed x $.5 per lb. = $150. If you compare that to the running total feed cost for the pig starter-grower in the previous chart, the difference is only $8. That's not bad.

Feeding Equipment Costs

There are two types of equipment used for feeding pigs on a homestead: (1) a feeding trough used for all forms of supplemental feed and (2) an automatic feeder used for pellet and granulated feed.

Feeding troughs will accommodate your desire to feed supplemental types of food, such as surplus or damaged fruits and vegetables, dairy by-products (like buttermilk and whey), table scraps, and lawn clippings. Wasted hay from goat and sheep feed bunks can be chopped up and fed with other forms of supplemental food.

Feeding troughs should be made of heavy metal to deter being turned over and can come in various lengths. A 2-ft. long metal feeding trough will run about $117, and a 4-ft. long version will cost about $130. You might need a 2-ft. version in both the boar's pen and the sow's pen and perhaps two 4-ft. versions in the feeder pig's pen.

Automatic feeders are the best and most economical way to distribute bagged feed, which can come in pellet or granulated form. The covered feeding troughs protect the feed from being spoiled by rain and from being wasted by the pigs.

These can come in single and double trough versions. A single version runs about $110, and the double version costs about $160. You would need a single version in the boar's pen and in the sow's pen. You might need at least four of the double versions in the feeder pig's pen.

Watering Equipment Costs

Watering equipment for pigs can consist of either an automatic watering nipple or an automatic hog watering cup. It is imperative that the pigs have adequate and accessible water. The automatic watering system ensures that the water is cool and clean.

You should have perhaps one automatic watering cup in each of the boar's, sow's, and feeder pig's pens. You can place the watering nipples at strategic places around the feeder pig's pen. Automatic water nipples are only $2–$3 each. A good cast iron automatic watering cup for hogs is about $130.

Containment Costs

Containment is made of hog panels and posts. An 8 ft. x 50 in. panel costs $25, and a 16 ft. x 48 in. panel will cost roughly $39. Heavy duty, 8-ft. T posts should be placed every 4 ft., and they cost about $9 each. Corner posts, 8 in. x 8 in., should be 8 ft. long, and they generally cost about $90–$110 each.

If you had two 8 ft. x 16 ft. pens, for the boar and the sow, and one 16 ft. x 40 ft. pen for the litter of feeder pigs, set side by side, and all three attached to a shelter on one end, the total containment setup for panels and post would cost $1116.

Shelter

The best shelter for hogs needs to be well-constructed and should be enclosed to provide shade when hot and shelter when the weather is harsh. The shelter will also be used for farrowing. With that in mind, the best shelter would be either a metal or wood barn.

Metal barns can cost $25–$50 per square foot, and wooden barns, on average, cost $45–$65 per square foot to build.

If you had a 12 ft. x 32 ft. shelter on one end of all three pens, the total square footage would be 384 sq. ft. That would allow for an inside pen 8 ft. deep and a nice walkway to have inside access to each pen—384 sq. ft. at $25 per sq. ft. would run about $9600.

Bedding

Bedding can be used for wintertime comfort and warmth and during farrowing. Bales of straw can be used at $5 per bale, or an 8-cu. ft. package of wood shavings can be used for $8.

During the winter, you could possibly use four bales per week. On average, you could spend $260–$320 per year for bedding.

Vet Care

Healthcare can be variable, but you should include at least an annual cost for dewormer and typical vaccinations for each pig.

Vaccinations are essential to prevent common diseases in hogs. The cost of vaccines can vary, but it might range from $2 to $10 per hog, and multiple doses may be required.

Hogs can be vulnerable to internal and external parasites. Deworming and parasite control measures can cost about $5–$15 per hog, depending on the products used.

Start-Up Expense Summary Charts

The following expense summaries are designed to provide a simple example for demonstrating start-up expenses. The following charts will be based on initial purchase of 1 boar, 1 sow, and the infrastructure necessary to raise two litters per year.

Initial Livestock Investment Expense Chart

Item	Cost
Boar (1)	$500
Sow (1)	$350
Total	**$850**

Infrastructure and Equipment Expense Chart

Item	Cost
Containment for boar, sow, and feeder pigs	$1116
Shelter and farrowing barn	$9600
Shipping container for feed storage	$2000
Feeding troughs for boar, sow, and feeder pig pens	$494
Automatic feeders for boar, sow, and feeder pigs	$860
Automatic hog watering cups	$390
Total	**$14,460**

Operating Expense Summary Chart

You will have to decide what kind of operation you want to have. The longer you feed the pigs, the more money you will have invested.

Feed Costs per Litter

Item	Retail Feed Cost for Litter @ 160 Lb.	Retail Feed Cost for Litter @ 250 Lb.
Boar feed for 6 mo.	$547	$547
Gestating sow	$268	$268
Lactating sow	$434	$434
Creep feed per piglet (14–40 lb.)	10.5 x 10 = $105	10.5 x 10 = $105
Pig starter per pig (40–60 lb.)	14 x 10 = $140	14 x 10 = $140
Pig starter-grower per pigs (60–100 lb.)	47.5 x 10 = $475	47.5 x 10 = $475
Pig grower per pig (100–160 lb.)	105 x 10 = $1050	105 x 10 = $1050
Pig finisher per pig (160–250 lb.)		223 x 10 = $2230
Total	$3019	$5249

Animal Care

Item	Annual Cost
Bedding	$300
Vet care @$225 per litter	$450
Total	$750

Revenue

There is a great opportunity to generate revenue from raising pigs because you can have two litters per year and have an average of 10 pigs per litter. That is, on average, 20 pigs per year you can raise and sell for meat.

In this example, one litter is sold at 160 lb. for meat, and one litter is sold at 250 lb. for meat. It is an interesting comparison. The following chart shows the typical feed cost for one litter of 10 piglets.

Item	Revenue for 160 Lb.	Revenue for 250 Lb.
Live weight	160 lb.	250 lb.
Hanging wt. (72%)	115 lb.	180 lb.
Hanging wt. price @ $4.75/lb.	$547	$855
Total revenue	$547 x 10 pigs = $5470	$855 x 10 pigs = $8550

Profit / Loss

Item	Expense (Sell @ 160 Lb.)	Expense (Sell @ 250 Lb.)	Profit/ Loss for 160 Lb.	Profit/ Loss for 250 Lb.
Total feed per litter (sell pigs @ 160 lb.)	$3019			
Total feed per litter (sell pigs @ 250 lb.)		$5249		
Annual animal care	$375	$375		
Meat processing cost	$75	$85		
Total expense	$3469	$5709		
Total revenue for litter @ 160 lb.			$5470	
Total revenue for litter @ 250 lb.				$8550
Total profit/loss			+ $2001	+ $2841

Profitability

You can increase your annual profitability by feeding both litters to 250 lb. and by feeding wholesale mill feed. The difference is worth the extra time and effort.

Profit with Different Litter Market Weights and Different Feed Costs

Item	Profit w/ Retail Feed	Profit w/ Wholesale Feed
Total profit for 160-lb. litter of 10	$2001	$3510
Total profit for 250-lb. litter of 10	$2841	$5465
Total annual profit	**$4842**	**$8975**

Profits for Same Litter Market Weights and Different Feed Costs

Item	Profit w/ Retail Feed	Profit w/ Wholesale Feed
Total profit for 250 lb. x two litters of 10	$2841 x 2 = $5682	$5465 x 2 = $10,930
Total annual profit	**$5682**	**$10,930**

Picture Gallery

Birkshire

Chester White

Duroc

Hampshire

Gloucestershire

Yorkshire

CHAPTER 6
Rabbits

R abbits are an amazing animal. They are a favorite on homesteads for raising meat, because they are so easy to raise and
provide a lot of meat with just one pair. In regard to convenience,
they are clean, quiet, and are neither messy nor smelly.

They are extremely prolific and relatively inexpensive to
feed and house. They are also a multipurpose animal in that
they can produce meat, ready-to-use manure, and fur pelts for
clothing. Rabbits are efficient meat producers. Just one pair can
produce 84 lb. of dressed-out meat per year with a moderate
breeding schedule.

In this chapter, breeding, kindling, feeding, watering, and
housing will be discussed. There will also be explanations about
meat production and processing, fur processing, and health
management. The chapter will close with a cost analysis and a
benefit/value appraisal.

Breeds

There are numerous domestic rabbit breeds that are used
in the US for meat and, to a lesser extent, for fur. The most

commonly used breeds for meat are listed below. After a brief description of each breed, charts listing desirable characteristics will be presented as a way of comparison.

At the conclusion of this section, there will be a brief mention of additional breeds that can be used for meat. In reality, any domestic rabbit breed can be used for meat. It is the performance in meat production efficiency that makes the difference.

Common Homestead Meat Rabbits

New Zealand White

The New Zealand White breed was developed in the United States. It is a large breed with erect ears. They have white fur, red eyes, and a medium-to-large build. They are easy to handle and have a high temperature tolerance. New Zealand White rabbits are known for their rapid growth rate, excellent dressing percentage, and good meat-to-bone ratio. They are the most commonly used breed in rabbit meat production.

Californian

This breed was developed in the United States. Their fur is white with distinctive black or dark brown areas on the nose, ears, feet, and tail. They are docile and have a high temperature tolerance. They are similar to New Zealand Whites in terms of growth rate and meat quality. They have a high dressing percentage and are a popular breed in commercial rabbit farming for meat production.

Flemish Giant

The Flemish Giant originated from Belgium. The rabbits are large in size and come in a variety of colors. Common colors include sandy, black, blue, and fawn. They have a gentle and friendly nature. They are also known for being generally hardy and cold-climate tolerant. Being a large breed, they are commonly raised for meat but are also renowned for their large appetite.

American Chinchilla

The American Chinchilla was developed in the United States. They have gray fur with a distinct pattern of darker and lighter bands. Chinchillas were originally bred for fur but are also raised for meat. They are medium in size and have a moderate litter size. They are generally docile and friendly and are typically tolerant of high temperatures. They have a good meat-to-bone ratio and dressing percentage.

Standard Rex

The Standard Rex breed originated from France. Their fur is short and plush and stands upright, giving a velvety appearance. They come in a variety of colors. Rex rabbits are raised for both meat and fur. The rabbits are medium in size and have a moderate litter size. They are generally calm and are typically tolerant of cold temperatures. The Rex has a moderate growth rate with good meat quality.

Satin

The Satin breed originated from the United States. Their fur is shiny and silky with a satin-like sheen. They come in a variety of colors. They are raised for their unique fur and are also used for meat production. They are medium in size and have a moderate litter size. They are calm and docile and are generally hardy. They have a moderate-to-fast growth rate and are efficient feed converters.

Silver Fox

The Silver Fox breed originated from the United States. Their fur is a dense, silver-tipped fur with a distinctive black colored undercoat. Silver Fox rabbits are known for their beautiful coats and are raised for both meat and fur. They have a calm temperament and are adaptable to different conditions. They have good growth rates and good meat quality.

Palomino

The Palomino breed was developed in the United States. They have golden or lynx-colored fur. Even though they were originally bred for fur, Palominos are also raised for meat production. They are medium in size and have a moderate litter size. They are generally hardy and are efficient feed converters. They have a moderate-to-fast growth rate and have good meat quality.

Cinnamon

The Cinnamon breed originated from the United States. They have rich cinnamon-colored fur. The Cinnamon rabbits are raised for both meat and fur. The Cinnamon rabbits are medium in size with a smaller litter size. They are adaptable to different conditions and are generally hardy. They have a moderate growth rate and good meat qualities.

Giant Chinchilla

The Giant Chinchilla breed was developed in the United States. This breed is similar to the American Chinchilla but is larger in size. They are raised for both meat and fur. They have a large litter size. They are adaptable and generally hardy. They are large in size with a moderate growth rate. The Giant Chinchillas have a moderate meat-to-bone ratio and a good dressing percentage.

Other Breeds

There are several other rabbit breeds that can also be used for meat production on homesteads: Standard Chinchilla, Champagne d'Argent, Crème d'Argent, Harlequin, Blanc de Hotot, Silver Marten, Florida White, Altex, Dutch, and Belgian Hare.

Breed Selection Considerations

Criteria	New Zealand	Californian	Flemish Giant	American Chinchilla	Standard Rex
Dressing percentage	Excellent	High	High	Good	Good
Meat-to-bone ratio	Excellent	Good	Good	Moderate	Moderate—Good
Growth rate	Rapid	Fast	Slow	Moderate	Moderate
Size	Medium	Medium	Large	Medium	Medium
Temperament	Docile	Docile	Friendly	Calm	Calm
Litter size	6–12	6–12	6–10	4–8	6–8
Adaptability	Tolerant to heat	Tolerant to heat	Tolerant to cold	Tolerant to heat	Tolerant to cold
Feed efficiency	Efficient	Efficient	Fair	Moderate	Moderate
Hardiness	Robust	Robust	Fair—Good	Hardy	Hardy

Criteria	Satin	Silver Fox	Palomino	Cinnamon	Giant Chinchilla
Dressing percentage	Good	Good	Good	Good	Good
Meat-to-bone ratio	Moderate —Good	Moderate	Moderate —Good	Moderate	Moderate
Growth rate	Moderate —Fast	Moderate	Moderate —Fast	Moderate	Moderate
Size	Medium	Medium	Medium	Medium	Large
Temperament	Calm— Docile	Friendly	Calm— Docile	Friendly	Calm
Litter size	4–6	6–8	5–7	4–6	7–10
Adaptability	Good	Tolerant to heat	Good	Good	Good
Feed efficiency	Efficient	Moderate	Efficient	Moderate	Moderate
Hardiness	Hardy	Hardy	Hardy	Hardy	Hardy

Gestation

The gestation period for rabbits is only about 30 days. This enables rabbits to exhibit a uniquely high reproductive rate compared to other livestock. They are also capable of re-mating soon after kindling (giving birth).

Having a short gestation period and being capable of mating relatively soon after having a litter make it theoretically possible to have as many as eleven litters per year. Of course, that is not healthy for the doe. Instead, a well-managed breeding program is recommended for controlling the timing of mating and the numbers of litters per year.

Breeding

If you are wanting to raise rabbits for meat, you will want to have as many litters per year as practicable but within reason. To maintain the health and vigor of your breeding stock, you should arrange for the litters to be well-spaced without overlap. Ideally, you should arrange for the doe to have a month's rest between litters.

Female rabbits (does) do not have a heat cycle like most other mammals. They are induced ovulators, which means that the doe is induced to ovulate by way of sexual activity, such as copulation. This makes it easy to formulate and follow a strategic breeding plan for mating, kindling, weaning, and remating.

Scheduling

It is possible for the doe to remate shortly after kindling, but it is not healthy for the doe to have an excessive number of litters per year. It is better to have a breeding schedule that is not excessively demanding on the doe.

Scheduling should take into account the gestation length and the weaning time of each litter. Litters are typically weaned at 6–8 weeks of age.

There are two commonly used schedules for meat producers—an aggressive or intensive schedule and a moderate schedule.

Aggressive Schedule

An aggressive schedule uses a 1-month breeding plan, when the doe is rebred about halfway through the 2-month litter period. This will yield about six litters per year.

Jan	Feb	Mar	Apr	May	Jun	Jul	Aug	Sep	Oct	Nov	Dec

wean   wean   wean   wean   wean   wean

Litter | 2-month Litter | 2-month Litter | 2-month Litter | 2-month Litter | 2-month Litter | 2-month

Mate   Kindle   Mate   Kindle   Mate   Kindle   Mate   Kindle   Mate   Kindle   Mate   Kindle

Moderate Schedule

A moderate schedule uses a 2-month breeding plan, when the doe is rebred at the end of the 2-month litter period. This will yield about four litters per year, with a 1-month rest between litters.

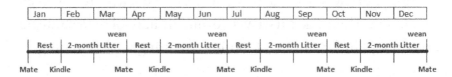

Jan	Feb	Mar	Apr	May	Jun	Jul	Aug	Sep	Oct	Nov	Dec

wean   wean   wean   wean

Rest | 2-month Litter | Rest | 2-month Litter | Rest | 2-month Litter | Rest | 2-month Litter

Mate   Kindle   Mate   Kindle   Mate   Kindle   Mate   Kindle   Mate

Most homesteaders who want to plan for breeding stock longevity generally follow the moderate schedule. If more kits per year are needed, more does can be added to the breeding stock.

Breeding Management

Separation

Because rabbits breed so readily, you will need to keep the buck separated from the does. This will allow you to manage your breeding program by controlling the mating dates, kindling dates, etc.

Health and Condition

Before you allow the rabbits to mate, you should verify their health and condition. You should allow the doe to be bred only if she is ready and able to kindle well and nurse another litter.

For the most part, assessing their health and condition is done by physical observation. You can also follow a health plan to ensure that they are free of the diseases and maladies listed later in this chapter under Health Management.

Placement

When it is time for the doe to mate again, you can place her in the cage with the buck. It usually doesn't take long for sexual activity to begin. At the conclusion of servicing the doe, the buck will curl up and fall to the side of the doe, signaling a successful copulation.

You may want to give the pair enough time to copulate at least twice in order to help ensure impregnation success. Even with two mountings, it only takes a few minutes.

Alternating Schedules

It is possible that you may want to have more kits per year than what 1 doe can provide. The best way to accomplish additional numbers is to add more does to your breeding stock, rather than using a more aggressive schedule.

When additional does are added to the program, you should arrange for their breeding schedule to be implemented into an alternating sequence with the others. For instance, if you had doe A on a moderate breeding schedule, wherein she was mated in January, you could arrange for doe B to enter on the same type of schedule but begin by mating her in February.

If you had a third doe C, she could be mated in March. Then in April, it would be time for doe A to be mated again and so on.

If you had 2 does on a moderate breeding schedule, you would have kits being weaned in the months of April and May, July and August, October and November, etc. If you had 3 does, you would have kits being weaned every month.

Once the kits are weaned, then they would be harvested as fryers at 8 weeks or enter into a finishing-feeding regime to be harvested at 12 weeks as roasters.

Kindling

Cage

The doe cage can be arranged many different ways. The types and dimensions of cages will be discussed in the Housing section later in the chapter.

The cage for the breeding doe is generally a wire cage that is constructed of hardware cloth. The general size should be large enough to provide enough space for the doe, her feeder, water source, and the nesting box.

Nesting

The nesting box size should be dictated by the size of the rabbit breed. The average size is 18 in. long x 10 in. wide x 9.5 in. high in the back and 5.5 in. high in the front. In brief, the nesting box is designed to house the kits as they develop, and they will remain in the box until they are about 4 weeks old.

On days 27–28 of the doe's pregnancy, you should prepare the nesting box by stuffing it with grass hay and placing it in the cage with the doe. When the doe is ready, she will begin to make her nest in the nesting box, by creating a space inside and lining the cavity with fur pulled from her stomach.

Note: If you place the nesting box in the doe's cage too early, she will more than likely eat the hay and might begin to use the nesting box as a litter box.

Kindling

The doe will generally kindle (give birth) somewhere between day 28 and day 31. You should check the nesting box each day during this time to see if she has kindled.

After kindling, you should check the nesting box and count the number of kits. Some people take them out of the nesting box as they conduct the count, so that they can verify their condition and also be able to remove any remaining placenta.

You should continue to check and count the kits each day for the first week. If a kit expires, you will need to remove the dead kit immediately to keep the nest healthy.

The doe will typically have 6–12 kits per litter. The average size litter is 8, and it is not uncommon for one of the kits to expire for a variety of reasons. These losses generally happen within the first week.

Litter Kits

The kits will be hairless when they are born. They will begin to grow fur their first week, and their eyes will begin to open in 10–14 days. Continue to check the nest each day to verify that the kits have plenty of insulation (fur). Add fur, when necessary, by pulling some off of the doe's stomach, and remove any soiled fur from the nest in order to keep it clean and free of germs.

The doe will nurse the kits about two times per day. Don't be surprised if you don't see this nursing activity. It's normal not to witness it. It doesn't take the doe long to nurse, and she may do this mostly at night.

At 2–3 weeks of age, the kits may jump out of the nest and not be able to get back in. This is normal. Simply place the kits back in the nesting box. Also make sure that the kit snuggling area is at the back of the box, so they have the best chance of staying warm.

Sometimes, the doe will eat part of the hay in the nesting box. If this happens, replenish the grass hay to ensure that the nest box interior has plenty of nesting material for insulation purposes.

At week 3, the kits will begin to drink some of the doe's water and eat some of her pellets. At week 4, they are able to navigate around the cage without the need of the nesting box. At week 4, you should be able to remove the nesting box from the cage.

Precautions

Hot summers and cold winters make kindling a challenge, and most homesteaders try to avoid kindling during these extreme temperature months.

These extreme temperatures are hard on both the doe and the kits. The litter sizes are generally smaller, and the kits generally take longer to grow and develop. There is also more chance of losses during these extreme temperature months. For that reason, you may want to avoid raising kits during summers and winters, unless these months are relatively mild in your location.

Weaning

The kits are typically ready to wean at 6–8 weeks, with most homesteaders weaning at 6 weeks. Weaning at 6 weeks allows the doe 2 additional weeks to recover before the next litter.

Weaning too early can cause a sudden change in the diet too abruptly and lead to serious digestive issues like enteritis and gastrointestinal stasis. More will be discussed on these issues in the Health Management section later in this chapter.

To wean the kits, they are removed from the doe's cage and placed in finishing cages or a finishing hutch. Finishing cages are generally kept in a barn or under a shed, in order to provide shade and protection from the weather. A finishing hutch can be kept outdoors, as the hutch has a protective space on one end.

Each kit will need roughly 2 sq. ft. of space. More detailed description and dimensions will be provided in the Housing section later in the chapter.

Once in their finishing cages or hutch, the kits should have free access to water and feed. The kits will not need to be separated by sex, because the general plan is to process the rabbits at either 8 or 12 weeks for meat. It they are kept beyond 12 weeks, the males will need to be separated from the does.

Feeding

In the following explanations for feeding, watering, and meat production, the New Zealand or Californian breed is used.

Adult rabbits will eat about 4 oz. a day, and does with young need about 8 oz. Meat rabbits, on average, will consume

1–1.5 oz. of feed per pound of body weight per day. This equates to about ½–1 cup a day, depending on age and breed.

Does

Breeding does will be housed in a cage constructed of hardware cloth. The cage should contain a feeder, a waterer, and at times, a nesting box.

Most of the time, the doe will either be pregnant or nursing and will thus need a well-balanced, nutritional diet. Rabbit pellets are designed to provide this need.

Both feed quality and quantity are important. When not nursing a litter, the doe should be fed roughly 1 cup of pellet rabbit feed per day (perhaps more in winter). The feed should consist of the following: 2–5% fat, 15–19% fiber, and 17–18% protein.

Once the doe kindles, she should continue to be fed 1 cup of feed per day, until the kits are 2 weeks old. At 2 weeks, the kits will begin to jump out of the nesting box and nibble the pellets. At that point, you should fill the cage feeder two times per day, in order to have free access to the feed for both the nursing doe and her kits.

This feeding regime will continue until the kits are weaned at 6–8 weeks of age. At that time, the kits will be removed and placed in their own finishing cages or a hutch. Once the kits are removed, the doe should go back to being fed 1 cup of feed each day.

Kits

The kits will begin to sample solid food at week 2–3 and will continue to make that transition the entire time they are with the doe. By week 6, quite a bit of their diet will be from eating pellets and a bit of grass hay. The entire time they are in the cage with the doe, they should have free access to pellet feed and a bit of grass hay, if available, as an alternative for fiber.

Once weaned, the kits will be moved to either finishing cages indoors or a hutch outdoors where they will be fed until they are ready to process for meat. The goal is to raise the kits to either a 5-lb. fryer or a 7-lb. roaster.

Indoor Finishing Cage Feeding

Finishing cages are constructed of hardware cloth and can either be elevated or stacked. Elevated cages allow the droppings and urine to fall to the ground underneath. Stacked cages need a tray underneath each cage to catch the droppings and urine.

Feeding in indoor finishing cages use either a J feeder or a stoneware bowl. J feeders usually have a top lid with a gravity feed tray bottom. These feeders are generally hung on the wall of the cage, with the container part on the outside, and the open trough extending through the wall to the inside of the cage, allowing free access.

The stoneware feed bowls are generally 3–5 inches in diameter and 2–3 inches high. They are designed to be heavy enough to avoid tipping and spilling.

The kits should be fed rabbit pellets that consist of the following: 2–5% fat, 15–19% fiber, and 15–16% protein. The feed should be formulated to optimize meat production.

In addition to the pellets, you can feed a bit of grass hay to assist in the need for fiber. Note that rabbits may nibble on food during the day but do most of their eating early in the morning and late in the afternoon. It is, therefore, important to have their feeders full of feed at these times.

Outdoor Finishing Hutch Feeding

The outdoor hutch can be stationary or it can be mobile. The mobile version can be picked up and moved to a grassy area. Both versions are constructed with the same dimensions.

Both hutch versions have basically the same dimensions and features. They are basically 2' wide x 5' long x 2' tall with a 2' x 2' enclosed space on one end. A description with dimensions will be discussed in more detail in the Housing section of this chapter.

Stationary Hutch

For stationary versions, the hutch is elevated to allow the manure to fall to the ground with enough space underneath for easy removal. The hutch should contain a feeder and a waterer, and the kits should have free access to pellet feed and a bit of grass hay.

Mobile Hutch (Tractor)

For mobile versions, the hutch tractor will be lifted up and moved to a new grassy area. The kits will be able to forage on the grass and also have access to feed pellets. If the grass is of good height and quality, it is possible for the kit's diet to consist 100% on the grass forage, but the rate of weight gain will be affected.

If mobile hutch tractors are used, it is imperative that they be moved to a new space every day for sanitary and health purposes. Please note that a large grassy area will be required to accommodate a daily moving plan.

The more forage percentage that makes up the diet, the longer the finishing time. If the kits are fed pellets only, it is possible to achieve 5 lb. at 8–10 weeks and 7 lb. at 12–14 weeks. However, if forage is used to supplement their diet, these goals can be extended an additional 2–4 weeks.

Please note that if forage is used to replace pellets altogether, the rabbit's nutritional needs may be lacking and affect the quality and quantity of meat. With that in mind, it is often recommended to offer free access to pellets to supplement the forage. Pellets will also help to keep their teeth ground down for they grow continually.

Watering

Providing water access to your rabbitry is quite simple, and the watering equipment usually consists of either a water bottle or a stoneware crock. Even though water bottles work just fine, the quantity of water can be somewhat limited during hot months when water consumption needs to increase.

The water consumption of meat rabbits can vary based on factors such as age, size, health, diet, and environmental conditions. On average, rabbits typically need about 1–2 oz. of water per pound of body weight per day.

The most important thing to remember is to keep cool, clean water available at all times during day and night. The best way to do this is to use a stoneware crock for plenty of water access and a water bottle as a backup, in the event the crock is depleted during hot days or accidentally spilled.

Stoneware crocks used for water are generally 5 inches in diameter and 3 inches tall. They should be refilled throughout each day whenever the water level is low or the water is dirty.

Water bottles used in a rabbitry are typically plastic and come in a variety of sizes. The typical sizes used for rabbit cages are in the 32–64 oz. range with the 32-oz. size the most common.

The bottles have a nipple on the dispensing end, which allows rabbits to access the water on demand. It is recommended to check the nipples from time to time to ensure that they are working properly and not clogged.

During the winter, you can use heated water bottles to prevent the water from freezing. This will be a significant time saver.

Meat Production

The goal for meat production is to achieve the desired live weights as quickly and efficiently as possible. The fastest and most productive method is to keep your rabbits on 17–18% pellets throughout their feeding regime.

The objective for most rabbitries is to arrive at the desired fryer live weight of 5 lb. at 8–10 weeks or a desired roaster live weight of 7 lb. in 12–14 weeks.

It is worth noting that the most economical weight for meat production is the fryer live weight of 5 lb. At this milestone (5 lb. at 8–10 weeks), the feed conversion ratio is roughly 3:1.

Thereafter, the ratio will go up rather steeply. For instance, it can take an additional 4 weeks of feeding just to gain 2 additional lb. for the roaster weight.

The dressing percentage for meat rabbits raised on pellets is generally about 60%. Fryers at 5 lb. live weight will tend to dress out at 3 lb. of meat, and roasters at 7 lb. live weight will dress out at 4 lb. of meat.

Meat Marketing

Before you decide to raise meat rabbits to sell, you will need to verify that you have a market for your rabbit meat. There are several places where you might be able to market your meat. If you don't have a viable option for marketing your meat, you can always optimize your home use.

Farmers' Markets

Many homesteaders sell their products, including rabbit meat, at local farmers' markets. These markets provide an opportunity to connect directly with consumers who are looking for locally raised, high-quality products.

Local Butcher Shops

Contacting local butcher shops or meat processors may be an option if you don't want to handle the processing and packaging of the rabbit meat yourself.

Food Coops

Local food cooperatives often work with small-scale producers to offer a variety of farm-fresh products to their members.

Local Homestead Network

Some homesteaders are able to sell directly to neighboring customers through a local farm customer network.

Home Use

Most homesteaders raise rabbits for their own meat. With a trio of 1 buck and 2 does, it is possible to raise 168 lb. of meat in 1 year with a moderate breeding schedule. Considering that rabbits have a 3:1 feed conversion ratio, and that the space for raising rabbits is minimal, 168 lb. per year, or 14 lb. per month, is significant.

Meat Processing

Processing your rabbit for meat is fairly simple. You will need to have a prepared processing station before you begin.

The processing area generally will include a method for dispatching; a method for hanging the rabbit for skinning and butchering; a bucket for the blood, entrails, and skin; a cleaning station; and a container of water or salt brine for soaking finished carcasses.

Dispatching

Dispatching (killing) the rabbit should be done by using a method that is quick, effective, and humane. The most common methods involve the dislocation of the head from the spine, causing the rabbit to expire instantly. The first thing to do after dispatching is to hang the rabbit and sever the head, so it can begin to bleed out.

Skinning

The rabbit should be hung upside down by small ropes or cables at shoulder height. The rope or cable should be attached to the back legs, just above the hock joint, and attached to a hanger that keeps the two back legs stretched apart.

Incisions are made at the hock, just below the rope or cable. The skin is then detached from the leg, all the way around at the leg—on both legs. The hide is then pulled down to the crotch area. Make a hole between the skin and the meat at the crotch area and cut the skin crosswise. This allows the skin to be pulled downward in front and back.

The skin is now pulled downward from the groin area in the front and pulled up and over the tail, and the tail is removed with clippers. This frees the skin so that it can be pulled downward in the front and back until you get to the front feet. Remove the feet with clippers, allowing the front feet and skin to be totally removed.

Butchering

Pinch the skin and make an incision above the stomach, a third of the way between the groin area and the front of the rib cavity. Put two fingers under the skin and slice the skin upward toward the groin area, careful not to cut the entrails. Grab the entrails above the bladder and pull downward until they are detached from the groin area.

Next, put two fingers under the skin at the stomach area and slice downward to the front of the rib cage, careful not to nick the intestines. You now have access to the body cavity. Reach in and remove all of the intestines and organs and drop them into the bucket.

Next, remove the diaphragm, heart, lungs, and esophagus. Some homesteaders like to keep the liver (with the gall bladder removed), the kidneys, and the heart for dog food.

If you want, you can then make an incision between the back legs, at the center of pelvic girdle, and apply pressure to break and open up the pelvic cavity. You can now remove the rabbit carcass from the hanger by snipping each hind leg at the hock joint.

Cleaning/Soaking

Once the butchering process is complete, you can take the carcass over to the wash station and clean the carcass of any blood or remaining body cavity tissue. If you are going to butcher more than 1 rabbit, you can soak the carcasses in a container of cold water. Some homesteaders prefer to soak the carcasses in a salt brine, which helps to tenderize the meat and prevent dryness when cooking.

Chilling

After all of the rabbits have been butchered and cleaned, the meat needs to be chilled in the refrigerator at 40 degrees or less for about 24 hours. The purpose for the chilling is to allow time for the meat to pass through the rigor mortis stage, which allows the meat to be more tender when cooked.

Fur

Raising rabbits for meat and raising them for fur is generally not mutually compatible because of ages. Typically, rabbits are processed for meat between 8 and 14 weeks for feed conversion efficiency and litter planning. Pelts from rabbits at 8–14 weeks old are typically not kept for fur because the hide is too thin and tears too easily.

Rabbits raised for pelts are normally not harvested until they reach a minimum of 4–6 months of age, when the pelt hide is thicker. The most common rabbit breeds for pelts are the Sable and the Chinchilla.

Tanning

There are several methods that can be used for tanning rabbit pelts. The following procedure is a basic home tanning process using salt and alum.

Prepare the Pelts

You should begin by removing any excess flesh, fat, or tissue from the skin. You can use a dull knife or scraper for this.

Salt Cure

Next, perform a salt cure by laying the pelts flat, fur side down, and generously sprinkle salt on the flesh side. Then roll up the pelts and store them in an airtight container in the refrigerator for 24–48 hours. This helps to draw out moisture and prevents bacterial growth.

Remove Salt and Rehydrate

After the salt cure, the pelts should be rinsed thoroughly to remove excess salt. Use a mild soap, if necessary, but rinse well to avoid leaving soap residue. After the salt removal, you should rehydrate the pelts by soaking them in lukewarm water for a few hours. This will tend to make the skin more pliable.

Tan the Pelts

Dissolve alum in warm water to create a tanning solution at a ratio of about 1 pound of alum per 1 gallon of water. Next, stir until the alum is completely dissolved.

Submerge the pelts in the tanning solution, and make sure the solution covers the pelts completely. Stir occasionally while letting them soak for about 2–3 days. Larger or thicker pelts may require longer soaking times.

Neutralize

Remove the pelts from the tanning solution, and then thoroughly rinse the pelts to remove any excess tanning solution. A mild baking soda solution (1 cup baking soda per 1 gallon of water) can be used to neutralize the pH. Soak the pelts in the solution for 15–20 minutes. After the neutralizing process, you should give the pelts one final rinse.

Stretch and Dry

Stretch the pelts to their natural shape and size, and tack them onto a wooden board or frame with the fur side down. Let the hide dry in a well-ventilated area away from direct sunlight.

Softening

Once dry, the pelts may be stiff. There are a couple of methods for softening the hide. The common methods are using a softening tool made of wood and using a softening oil.

Wood Tool Method

One method for softening is to attach a shortened wooden broom handle with a rounded top to a chair. As you are sitting, place the pelt, leather side down, on top of the broom handle. As you are pulling down on each side, oscillate the pelt back and forth over the top of the broom handle.

Continue this process over the entire area of the pelt until the hide has been softened. This process breaks the leather fibers, causing the hide to be soft and pliable.

Softening Oil Method

Another method for softening is by using a thin oil. Neatsfoot oil is a commonly used substance to soften hides after tanning. It helps to lubricate the leather, making the hide more pliable and supple.

While using a clean cloth or sponge, moderately apply neatsfoot oil to the entire surface of the hide. Be careful not to apply too much. Once finished, wipe off any excess, and let it sit overnight to allow absorption.

Allow the hide to air dry completely. As it dries, continue to work and flex the hide periodically to maintain its flexibility.

Brush the fur to restore its natural appearance and remove any tangles. Your rabbit pelts are now tanned and ready for use.

Marketing

Rabbit fur is soft and pliable making it a favorite for clothing like hats, coats, and gloves. Over the years, synthetic options have caused rabbit fur to be in less demand, resulting in lower pelt prices. Even though demand is not what it once was, some markets for rabbit fur still exist.

Typical prices for rabbit fur are about $10–$30 per pelt, depending on size and quality. The breeds that bring the highest prices are Sable and Chinchilla rabbits.

Selling rabbit pelts as a homesteader can be done through various channels, both online and offline. Here are some options.

Local Farmers' Markets

Many local farmers' markets allow homesteaders to sell their products, including rabbit pelts.

Craft Fairs and Artisan Markets

Craft fairs and artisan markets often attract individuals interested in handmade and natural products.

Online Platforms

- Etsy
- eBay
- Facebook Marketplace
- Amazon

Even though it is possible to tan and sell your rabbit pelts, the enterprise should be a by-product of another primary goal, such as meat, rather than being the sole objective. For instance, the cost of feed for raising a rabbit to 4–6 months is going to be far more expensive than what the pelt can yield in revenue.

Also, even if you do find a market to sell your pelts online, you should be prepared to find a lot of competition. Another option would be to make a product, such as clothing, from the fur and sell the product.

Housing

Unlike other forms of livestock, rabbits can be raised in cages or hutches, which is a huge space saver. Cages are generally made from hardware cloth and welded wire with either a wood or metal frame.

Wire cages are durable and tend to be more cost-effective over time. Wire cages are able to facilitate cleanliness, as the manure can fall through the wire onto the ground, and the wire is relatively easy to clean and disinfect.

The most common approach is to have the breeding and kindling cages indoors and the litter cages either indoors as grow-out cages or outdoors in a hutch or tractor.

Breeding Doe and Buck Cages

The cages are generally housed in a large enclosure, such as a shed or barn. The most common and effective arrangement is to have the cages elevated so that the cage floor is about 3.5 ft. to 4 ft. off the ground.

To elevate the cages, they are typically supported on stands or hung from the ceiling of the structure. To save space, the cages can be situated end to end.

Construction

Rabbit cages can be homemade or purchased as prefabricated. It is important to plan for the doors to be wide enough and tall enough to accommodate the insertion and removal of the nesting box.

The dimensions of the breeding cages are roughly 30 in. wide x 36 in. long x 18 in. tall. The frames can be either wood or metal.

The cage sides and tops can be constructed with welded wire or hardware cloth. If hardware cloth is used, the type should be ½ in. hole size and at least 19 gauge.

The bottom flooring should be constructed with 14 or 16 gauge welded wire with the hole size of ½ inch x 1 inch. The side with the wires ½ inch apart should be on the inside of the cage floor facing upward. This helps with weight distribution of the rabbit, which will assist in preventing sores on the rabbits' hocks.

Each cage should have a feed source, a water source, a rubber resting mat, and a nesting box for the breeding does.

Ventilation

Your cages and rearing area need to be well-ventilated. Adequate airflow is needed to void the rearing space of concentrated fumes from collected manure and urine.

More important, the cages need adequate airflow for cooling purposes—especially during hot summer days. Because the rabbits have fur coats, they can withstand cold weather a lot better than they can hot weather. They will suffer without adequate cooling on hot days.

Manure Management

The rabbit droppings and urine will fall through the openings in the cage floor wire and collect underneath. It is a common practice to manage this by regularly removing this manure to prevent flies and strong odors.

The manure can be added to a compost pile, which can later be used in the garden to amend and add nutrients to the soil. The manure can also be used right away for plants and in the garden without being composted. Rabbit manure is not too hot in nitrogen as are other types of livestock manure.

Grow-Out Cages

Once the litter kits get to their weaning age (6–8 weeks), they need to be transferred to a different cage where they will grow out to harvesting weights. The kits are normally placed in a grow-out cage together as a litter.

Indoor Cages

Grow-out cages for indoors are similar to the breeding cages. The dimensions are roughly 30 inches square and 24 inches tall for adequate air circulation.

The cages can be elevated like the breeder cages, or they can be stacked in two or three tiers. If they are stacked, each cage would be equipped with a tray underneath to catch the droppings and urine. The manure trays would need to be removed, dumped, and brushed off every day.

Even though stacked cages allow for more effective space utilization, the rabbits will be relatively close to their droppings, making it more possible for respiratory issues.

With this in mind, it is important to dump the manure trays daily and have an adequate ventilation system in place—especially in the summer. Electric fans can greatly assist toward the air flow needs.

Outdoor Hutch/Tractor

Outdoor cages are generally larger, requiring a sturdy frame. These large, outdoor cages are called hutches. They typically are constructed with wooden frames and covered with wire on the sides, for adequate ventilation, and on the bottom, for manure management.

Construction

The tops can be made of a solid material to provide shade and protection from weather. The sides are covered with welded wire or hardware cloth, and the bottom needs to be 14 or 16 gauge welded wire flooring, with ½ in. x 1 in. holes.

The dimensions can vary, but a popular size is 2 ft. wide, 5 ft. long, and at least 2 ft. tall. On one end, a 2 ft. square room with solid wood walls is constructed (generally made of plywood) with an opening to the open cage area.

In essence, the hutch provides a 2 ft. x 3 ft. wired cage area for roaming, stretching, and eating, and an adjoining 2 ft. x 2 ft. walled-in room at one end as a retreat and protection from harsh weather.

The top is generally hinged, allowing the top to be opened for easy access for rabbit management and feed and water maintenance.

Permanent or Mobile

The hutch can be permanently located and elevated on bricks or it can be mobile. Mobile versions simply have handles on each end, allowing the hutch to be lifted and placed on a grassy surface.

The mobile version is often called a rabbit tractor. The hutch can be laid on a grassy surface to allow the rabbits to forage, but it is important to move the hutch each day to a fresh grassy spot for sanitary purposes.

Health Management

Rabbit health is essential for a successful rabbitry. It is, therefore, important to be aware of common health issues that can affect your meat rabbits. Some of the most common health issues are listed below.

Gastrointestinal Stasis

This is a condition when the rabbit's digestive system slows or stops functioning properly. In young rabbits, it is generally caused by inadequate fiber in the diet. Common symptoms include little or no poop and a lack of appetite. Common remedies include the increase of fiber in the diet (such as hay), adequate water supply, and stress reduction.

Respiratory Infections

Damp or poorly ventilated conditions can cause respiratory issues in your rabbits. Typical symptoms may include sneezing, runny noses, labored breathing, and lethargy. Remedies include improved ventilation in the rabbitry and keeping the rearing areas dry and clean.

Enteritis

Enteritis is an intestinal inflammation, which is often caused by sudden dietary changes, such as eating too many pellets when first weaned. Reduced appetite, diarrhea, and lethargy are common signs. Some have suggested that offering Mylicon anti-gas drops can help as well as ensuring a consistent diet. You should consult a veterinarian if problems persist.

Rabbit Hemorrhagic Disease (RHD)

This is a viral disease that can spread quickly through a rabbitry. Symptoms include high fever, lack of appetite, lethargy, respiratory problems, seizures, nasal bleeding, and the illness can lead to sudden death. If you suspect an outbreak, you should consult a veterinarian. Vaccinations are the most effective way to prevent RHD.

Pasteurellosis

This is a bacterial infection that can lead to respiratory problems. Symptoms can include difficulty breathing, runny nose, red or watery eyes, and/or sneezing. Remedies often include isolating infected rabbits and the consultation of a veterinarian for antibiotic treatment.

Ear Mites

Ear mites can cause damage in a rabbit's ears. This will often lead to itching and shaking of the head back and forth. Remedies may include ear drops and perhaps medications that contain ivermectin or other parasiticides. You should follow your vet's instructions carefully when applying these medications.

Sore Hocks

Sore hocks occur when the skin on the rabbit's hind feet becomes damaged or injured from the wire floor. Ensuring the floor is made from 14 or 16 gauge welded wire with hole spaces of ½ in. x 1 in. will help to distribute the weight better. The wires creating the ½ in. spacing should also be facing upward inside the cage. Rubber rest mats will also help to give the rabbits a place to lie off of the wire flooring.

Heat Stress

Being sensitive to high temperatures, rabbits can suffer from heat stress. Be sure the rabbits have plenty of shade, cool water, and good ventilation during hot weather.

Dental Issues

Rabbits' teeth grow continuously, and they avoid overgrown teeth by their normal eating and chewing routine. By feeding your rabbits pellets, some grass hay, and perhaps having chew toys available in their cages, you can prevent rabbits' dental issues.

Parasites

External parasites, like fleas and ticks, and internal parasites, like coccidia, can affect rabbits. You should follow a regular health maintenance plan for your rabbits by deworming and inspecting for external parasites regularly.

In order to catch and address health issues early, you should follow a routine health monitoring program by regularly inspecting your rabbits for any signs of illness or discomfort.

Expenses

Start-Up Expense Summary Charts

Start-up expenses will include the initial investment for a breeding trio (1 buck and 2 does) and the cost for housing and equipment.

Initial Livestock Investment Expense Chart

Item	Cost
Price for trio (1 buck & 2 does)	$150

Infrastructure and Equipment Expense Chart

Item	Cost
Breeding/nesting cages	3 @ $45 ea. = $135
Hutch	$200
Feeders (1 per cage, 2 in hutch)	5 @ $9 ea. = $45
Water crocks	5 @ $8 ea. = $40
Water bottles	5 @ $7 ea. = $35
Nesting boxes	$23
Resting mats	5 @ $9 ea. = $45
Total	**$523**

Operating Expenses

The operating expenses are based on maintenance per litter, which will include feeding and animal care. The following figures will be based on a litter of 7, harvested for meat production as fryers at 5 lb. live weight each.

Feed Costs

Feed consumption per rabbit is about 1–1.5 oz. per day per lb. of body weight. The easiest method to determine the total amount of feed consumed is by using the feed conversion ratio of 3:1.

The fryers are harvested at 5 lb., so each kit will have consumed roughly 15 lb. of feed by the time they arrive at a 5-lb. fryer weight.

The doe will consume roughly 8 oz. per day for 6 weeks. The feed cost for the buck will be divided between 2 does.

Feed Cost per Litter

Item	Total Feed	Price per Lb.	Total Cost	Total for Litter of 7
Pellets—lb./doe	21 lb.	$.48	$10.08	$10.08
Pellets—lb./buck	10.5 lb.	$.48	$5.04	$5.04
Pellets—lb./kit	15 lb.	$.48	$7.20	$50.40
Total	N/A	N/A	N/A	$65.52

Animal Care

Item	Cost per Litter
Bedding—½ bale grass hay	$2.50
Vet care	$2.50
Total	$5

Cost Analysis

The following analysis compares the total cost per litter to the total amount of meat produced per litter in order to arrive at the meat cost per pound.

Item	Expense	Meat	Cost per Pound
Feed per litter	$65.52		
Care per litter	$5.00		
Total expense	$70.52		
Meat production/litter		21 lb.	
Cost per lb. of meat			$3.35

Benefit/Value

The following chart shows the total amount of meat per kit and per litter, meat per doe annually, and meat for 2 does annually. This is the annual benefit realized by a trio of 1 buck and 2 does. As shown in the chart above, the meat cost is roughly $3.35 per pound of meat.

Item	Total Live Wt.	Total Processed Wt.
Meat per kit	5 lb.	3 lb.
Meat per litter	35 lb.	21 lb.
Meat per year, per doe (4 litters)	140 lb.	84 lb.
Total benefit/value per year (2 does)		**168 lb.**

The market price per pound for rabbit meat varies greatly. Prices mentioned online range from $1.50 per lb. to $10 per lb. This largely is dependent on local demand. Because rabbit meat is not a product typically sold in a meat market or grocery store, it is difficult to know if you will be able to sell your rabbit meat, and if so, for how much.

The best way to determine the value of your rabbit meat is to compare the cost per pound to another type of meat you would be purchasing for your own household consumption if you were not raising your own meat. Compared to other typical meat prices, rabbit is going to be more than chicken, but less than beef or pork per pound.

For homesteaders wanting to be self-sustaining, raising rabbits for meat is a simple, cost-effective solution. Please also note that the price per pound for your rabbit meat can be reduced by foraging, but it takes longer to harvest.

Picture Gallery

New Zealand

California

Flemish Giant

American Chinchilla

Standard Rex

Satin

Silver Fox

Palamino

Cinnamon

Giant Chinchilla

PART 2

Poultry

Part 2 covers chickens, ducks, geese, and turkeys. The various breeds most commonly used to accomplish the goals for eggs and meat will be discussed. Some poultry have additional side benefits, such as weed control.

Each chapter covers management techniques for breeding, feeding, watering, and foraging management. Important requirements for containment and shelter are also discussed. Key elements, such as egg production, meat production, and meat processing, will also be explained in detail.

Each chapter lists common health issues and the importance of health management. At the end of each chapter a cost analysis and a benefit/value assessment will be provided. Finally, each chapter will conclude with a picture gallery of images for the referenced breeds.

Part 2 is full of vital information to help the reader fully understand what is involved in raising poultry in order to be self-sufficient. You will discover which type of poultry will best meet your needs, and which breed will be the best fit for your homestead.

CHAPTER 1

Chickens

Homesteads wanting to be self-sustaining and raise their own food should have chickens at the top of the list for farm animals. Chickens are amazing creatures. They are easy to raise, and, with little investment, they can give you daily eggs and occasional meat for the pot.

This chapter will compare the most common chicken breeds used for eggs and meat. It will discuss how to start by incubating eggs or start by raising chicks in a brooder. From there, how to achieve the best results through the various stages will be covered.

Flock maintenance, such as water and feed equipment and ration requirements, will be explored, as well as typical health requirements. Basic shelter and containment essentials, such as chicken coops and chicken runs, will also be described.

This chapter will conclude with a cost/benefit analysis in order to assist you in the decision-making process.

Breeds

There are many varieties of chickens to choose from in the U.S. for homesteads. Some breeds are used primarily for eggs, while others are used primarily for meat.

Breeds for Eggs

The most common 7 breeds used for eggs are listed below. Even though these breeds are mostly used for eggs, they can also serve a dual purpose for meat.

Rhode Island Red

Rhode Island Reds are known for their deep red plumage and excellent egg-laying abilities. They are medium-to-large-sized birds originating from the United States. They are docile and friendly, making them great for backyard flocks. The Rhode Island Red can produce approximately 200–300 brown eggs per year.

Leghorn

Leghorns are small-to-medium-sized chickens known for their prolific egg production. They typically have white plumage but come in various varieties. Leghorns originated in Italy and are highly active and alert birds. The Leghorn can produce approximately 280–320 white eggs per year.

Sussex

Sussex chickens come in a variety of colors, including red, white, and speckled. They are medium-sized birds with a friendly and gentle temperament. Sussex chickens originated in England and are valued for their dual-purpose nature, both for meat and eggs. The Sussex can produce approximately 250–275 brown eggs per year.

Plymouth Rock

Plymouth Rocks, also known as Barred Rocks, have distinctive black and white striped plumage. They are medium-to-large-sized birds originating from the United States. Plymouth Rocks are known for their calm disposition and excellent egg production. They can produce approximately 200–280 brown eggs per year.

Australorp

Australorps are medium-to-large-sized chickens with black plumage. They hail from Australia and hold the world record for the most eggs laid by a single hen in a year. They are known for being friendly and easy to handle. Australorps can produce approximately 250–300 brown eggs per year.

New Hampshire Red

New Hampshire Reds are medium-to-large-sized chickens with deep reddish-brown plumage. They originated in the United States and are good dual-purpose birds for meat and eggs. They have a calm and friendly temperament. New Hampshire Reds can produce approximately 200–280 brown eggs per year.

Wyandotte

Wyandottes come in various color varieties, such as silver laced and gold laced. They are medium-sized chickens with a round and sturdy appearance. Wyandottes are known for their hardiness and adaptability to different climates. The Wyandotte originates from the United States and can produce approximately 200–250 brown eggs per year.

Breed Comparison for Eggs

Criteria	Rhode Island Red	Leghorn	Sussex	Plymouth Rock	Austra-lorp	New Hamp Red	Wyan-dotte
Origin	US	Italy	England	US	Australia	US	US
Color	Red	White	Red/Wht Speckled	Black & White	Black	Reddish Brown	Silver Lace
Size	Med—Large	Small—Med	Med	Med	Large	Med—Large	Med
Eggs per week	5	5–6	3–4	3–4	5–6	3	4–5
Egg size and color	Large Brown	Large White	Large Brown	Large Brown	Large Brown	Large Brown	Med-Lrg Brown

Breeds for Meat

Almost any chicken breed can be used for meat, but some breeds are more commonly used for their efficiency in converting feed to meat, and the quality of the meat.

The breeds more commonly used for meat are: Cornish Cross (white), Delaware, Freedom Rangers, Jersey Giant, and Orpington.

Cornish Cross (White)

The Cornish Cross originated from the United States. Their plumage is typically white in color. They are a medium-sized chicken and are renowned for their fast growth rate. Cornish Cross chickens are primarily bred for their meat production. They have tender, juicy, and flavorful meat, making them a popular choice for meat production in the poultry industry.

Delaware

The Delaware chicken breed was developed in the state of Delaware in the United States. They have white feathers with black barring on the neck and tail. Delaware chickens are considered a medium-sized breed and are primarily raised for their excellent meat quality. They are known for producing tender, flavorful, and juicy meat. The combination of their good feed conversion ratio and meat quality makes them a popular choice on homesteads.

Freedom Rangers

Freedom Rangers were developed in the United States and come in a range of colors, including red with black plumage. They are medium-sized and are bred for both meat and outdoor foraging. They have good meat quality with a balance of flavor and tenderness. They are known for their ability to thrive in free-range environments.

Jersey Giant

Jersey Giants were developed in the United States. They have black, white, or blue plumage. These chickens are among the largest poultry breeds. Jersey Giants are valued for their large size and meat production. While their meat is flavorful, it may not be as tender as some other breeds due to their size.

Orpington

Orpington chickens originated in England. Orpingtons come in various colors, including buff, black, blue, and white. Most are a copper color. They are a medium-to-large breed. Their meat is considered good with a balance of flavor and tenderness. Orpingtons are considered a versatile choice for homesteads due to their dual-purpose capability.

Breed Comparison for Meat

Criteria	Cornish Cross	Delaware	Freedom Ranger	Jersey Giant	Orping- ton
Weeks to process	6–8	12–13	9–11	16–22	16–20
Process wt. (lb.)	4–6	5–6	5–6	10–12	8–10

Mating

Mating between a rooster and a chicken usually begins with the rooster's courtship behavior, commonly referred to as a mating dance. The mating dance consists of several things performed by the rooster.

The rooster will typically puff up his chest, neck, and tail feathers in order to appear large and attractive. He will then strut around the hen with a proud gait to display his plumage. The rooster will also typically lower the inside wing and cluck or crow while circling the hen.

If the hen is receptive to the courtship, she will squat down low to the ground, allowing the rooster to mount her. The rooster will mount from behind and crouch low on her back. The rooster will often grab the top of the hen's head or her neck with his beak.

As the hen leans forward, the rooster will lean backward to accommodate the mating process. Both the rooster and hen have a similar orifice called a cloaca. The rooster will press his cloaca against the hen's cloaca. During this momentary cloaca kiss, the rooster will deposit semen and the hen will receive the deposit.

The sperm will then travel up the hen's reproductive track and will be stored in storage tubules in the oviduct. The sperm will remain viable for 7–10 days. Sperm will be released from the oviduct to fertilize eggs, as they are laid over the next 1–2 weeks.

Starting with Eggs

Whether you are starting your first flock of laying hens for the first time or growing your own replacements, doing so by hatching your own eggs can be exciting and fun. If you already have laying hens, one option, of course, is to allow 1 or more hens to sit on their own clutch of eggs. However, to have more control over numbers, timing, and success, using an incubator is a more common approach.

Egg Source

If you use an incubator, you will need a source of eggs that are reliable. If you have a rooster, and have been documenting mating activity, you can use your own eggs from a hen that is brooding. Other viable options include purchasing fertilized eggs from a reputable local farm, or you can order fertilized eggs from an egg hatchery and have them shipped to you.

Incubators

Incubators come in a number of sizes and sophistication. The best choice is to find one that has heat control, humidity control, an automatic timer, and a fan to facilitate heat distribution.

More recent models display heat and humidity digitally, and some models even have built-in candlers.

Incubation Process

If you have an incubator that has been used before, you will need to prepare the incubator before beginning the incubation process. To sanitize the interior, wash with a warm 10% bleach solution, followed by soapy water, then rinsed and dried.

Once sterilized, make sure the water reservoir is full, and you should turn it on to the beginning temperature and humidity to ensure that it will reach and maintain those parameters. Now you are ready for the first batch of eggs.

When you have the eggs and are ready to begin, carefully place the eggs in the incubator, and set the temperature and humidity. Chicken eggs take 21 days to hatch. During the first 18 days, the temperature and humidity should remain constant, and the eggs should be turned. From days 19 to 21, the temperature and humidity will change, and the daily turning will cease.

Days 1–18

The temperature should be set at 99.5–100.5 degrees F., and the humidity should be set at 40–55%. The eggs need to be turned daily—three times per day is minimum and five times per day is optimum.

If your incubator doesn't have an automatic egg turner, you will need to turn the eggs manually. The best procedure is to mark one side of the eggs with an X and mark the other side with an O. This will help to verify that all the eggs have been turned each time.

The purpose for egg turning is to prevent the chick from sticking to the shell. The embryo rests on top of the yolk, and, if not turned frequently, will get stuck between the yolk and the shell. By turning the eggs, the yolk will float through the albumen (egg white) to the top of the egg, and thus avoid the sticking-to-the-shell issue.

Days 7–10

Toward the middle of the first 18 days, you can candle the eggs to verify if the eggs are fertile and are progressing properly. Candling is a process of shining a light through the egg so that you can see the status of the egg. The most common method is to use a flashlight.

Candling should be done in a small, dark room. You should begin by removing only a few eggs from the incubator at a time, and they shouldn't be outside of the incubator for more than 5–10 minutes.

Shine the light through the egg. If the egg is clear and free of dark areas, the egg is infertile, and should be removed and discarded. If a ring of red is visible, there was more than likely an embryo initially, but it has died. The egg should be removed and discarded. If you see blood vessels within the egg, there is a live embryo inside. The egg should be returned to the incubator. You can repeat this process, a few eggs at a time, until all of the eggs have been checked.

Day 19

On the nineteenth day, egg turning should stop, and all the eggs should be placed so that the large end of the egg is at the top. The temperature should continue to be 99.5–100.5 degrees

F., but the humidity should be increased to 60–70%. Do not open the incubator until day 21.

Day 21

The eggs should begin to hatch on day 21. If some haven't, there is still time. It can take up to 24 hours for the pipping and resulting hatching to be complete. You should allow the eggs to hatch on their own. If you attempt to assist in this process, you can cause excessive and possibly fatal bleeding.

Once the hatching is complete, you should lower the temperature to 95 degrees F. Once the chicks are all dry, you can remove them from the incubator, and place them in the brooder.

Starting with Chicks

Small chicks can be purchased from reliable hatcheries, and you can have your order shipped to you. Orders from hatcheries can be placed for all females, all males, or mixed run, which means an unknown mixture of sexes.

Whether you hatch the eggs yourself with an incubator or purchase the chicks from an outside source, the new chicks will need to be moved to a brooder. The brooder provides a warm, safe environment where the chicks will be cared for until they are roughly 6 weeks old.

The brooder can be commercial or homemade. Homemade brooders can be constructed from a variety of materials. The size and material choice largely depends on the number of chicks being raised at any one time.

Homemade construction options can be something as simple as a carboard box or can be constructed out of wood, as a more long-term option. The most common options for homemade brooders are a cardboard box, a plastic storage container, a galvanized water trough, and a wooden box made out of plywood.

Regardless of the material choice, the brooder space should include a heat source, thermometer, bedding, feeder, and waterer. These options are discussed more fully in this section.

Brooder Options

Brooder size and material choices are often influenced by the number of chicks to be raised in the brooder and the number of batches planned.

Brooder Size

For space requirements, you should plan on having ½ sq. ft. per chick. For instance, a brooder that is 2 ft. x 4 ft. (8 sq. ft.) could comfortably accommodate 16 chicks. A brooder that is 4 ft. x 4 ft., could accommodate 32 chicks.

Cardboard Box

Cardboard boxes are sometimes used for one-time, small batches of one dozen or fewer. The best cardboard box choices are those that have walls with double thickness for stability. The box should be large enough to provide adequate space for the chicks and all of the equipment. The wall height should be roughly 24–30 inches tall to provide protection and security.

In order to find a box large enough, the best boxes are those that were used to ship large appliances like a refrigerator or stove. Cardboard is absorbent, so you should only plan on using the box for one batch.

Plastic Storage Container

Plastic storage containers make good brooders for small batches. Colored containers are preferred over clear ones in order to have better control with lighting. The best dimensions are those that are about 2 feet tall and 2 feet wide. The amount of depth will be determined by the number of chicks you plan to brood. The more common versions are 2 ft. wide x 2 ft. tall x 3–4 ft. long. These containers are often a preferred option because they come with a top that can be modified with a screen in the center, which allows ventilation and protection from rodents and pets.

Galvanized Water Trough

This is a more long-term option used by those who intend to brood more than one batch. The galvanized livestock water troughs come in a variety of sizes. You can choose one that will fit the number of chicks you plan to brood at one time. The most common sizes are those that are 2 ft. tall x 2 ft. wide x 4 ft. long. You will need to construct a top to accommodate protection but also allows for heat sources and easy access.

Wooden Box

Wooden brooder boxes constructed with plywood are another common construction option for long-term use. This option is sometimes preferred over the galvanized water trough option due to the cost savings.

Plywood typically comes in 4 ft. x 8 ft. sheets, so the most common size choice is a 4 ft. x 4 ft. square box with a wooden floor and custom-made top. Plywood sheets that are ½-inch thick should work the best.

These wooden brooders can be as simple or as complex as you want, depending on your future plans, size of broods, etc. There are a variety of construction plans that can be found online. Tops can be constructed by making a wooden frame made with 2" x 4"s and covering the frame with chicken wire. Hinges can be added for easy access.

Brooder Equipment and Practice

Every brooder will need the same type of equipment. The size and styles can vary. Each brooder will need a heat source, bedding, a waterer, and a feeder. In addition, best brooder practices should be followed for optimum growth and development.

Bedding

There are quite a few options for bedding material and methods of use. The most common material options include pine or aspen shavings, hemp shavings, coarse sand, and straw. Materials that are not recommended include cedar and teak shavings (because of toxicity issues), fine sand, and sawdust (because of dust issues).

For pine shavings, you should arrange the depth to be 3–4 inches thick and replace it once per week. The used shavings can be mixed with compost to be used in the garden.

When using coarse sand, the depth should be 3–4 inches. Droppings can be cleaned out daily with a scooper, like used for cat litter. It is recommended not to use heat lamps close to the sand, as it can make the sand too hot. A heat plate will work OK.

Straw can be used for bedding, as long as it is clean and not too chaffy or dusty. The layer should be 4–5 inches thick—thick enough to be absorbent and provide a layer of insulation.

The following chart compares pine shavings, sand, and straw for several key criteria.

Criteria	Shavings	Sand	Straw
Absorption	Good	Poor	Excellent
Cleanliness	Medium	Excellent	Poor
Bacteria hold	Med—High	Low	Med—High
Dust	Med—High	Low	Med—High
Insulation	Warm—3"–4"	Cool	Warm—4"–5"
Composting	Excellent	Poor	Excellent
Overall ranking	1	3	2

Heat Source

Heat lamps and heat plates are the most common options. Heat lamps commonly come with 250-watt bulbs. Temperature is controlled by either raising or lowering the lamp.

The use of heat plates is a relatively new heat option for brooders and has many advantages. The plate can be purchased in a variety of sizes, ranging from 12" x 12" to 16" x 24". They come with adjustable legs for height adjustment and a protective top. Temperature control regulators can be purchased separately. Many believe that these heat plates are safer, are easier to regulate temperature, and allow for a regimen of the timing and length of light and dark hours.

Heat Management

Your chicks will need heat 24 hours per day while in the brooder. The first week, the temperature should be 95 degrees F. and be reduced 5 degrees each subsequent week.

If you are using one or more heat lamps, you can adjust the temperature by raising or lowering the lamp(s). For a heat plate, you can purchase separately a temperature controller. You should have a thermometer under the heat source in order to verify the temperature. The best placement of the heat source is in the middle of the brooder with the waterer and the feeder on opposite ends of the brooder.

Another key way to tell if your heat source is adjusted correctly is to observe the chick's behavior. If the chicks are huddled together under the heat source, they are too cold. If the chicks are spread out all around the perimeter of the brooder, they are too hot. If they are huddled over to one side of the brooder, you may have a cold draft they are trying to get away from. The proper behavior for the correct temperature setting is when the chicks are evenly spread throughout the brooder space.

As mentioned before, the two most commonly used sources for heat are heat lamps and heat plates. One advantage of the heat plate is that it provides heat without light. This allows you to be able to control the number of dark and light hours each day.

Light Management

Even though it isn't mandatory, there are many advantages to using a heat source that allows you to regulate a day/night cycle for your chicks. Establishing and maintaining a controlled light schedule can help to reduce stress and optimize their growth and development.

If your goal is to raise hens for egg production, following a consistent light regimen can influence their laying patterns when they reach maturity. An often-suggested schedule begins with 23 hours of light for the first week and then reduce the amount of daylight by 2 hours each subsequent week. This process will allow you to arrive at 15 hours of daylight and 9 hours of darkness at week 5, mimicking natural daylight patterns.

Feeders

Self-feeding feeders for chicks have a saucer or bowl-shaped base with a screw-on container. Other types are small troughs with holes in the upper lid for easy access. Feeder selections should be a free choice feeder, allowing continual access.

The chicks will begin to eat the chick feed from the feeder right away. You should position the feeder so that the chicks can have access on all sides. The chicks will tend to hop and scratch out some of the feed, which gets mixed with bedding and wasted. To mitigate this, you can place the feeder on top of a ½" board.

In the first 6 weeks of growth, chicks will double their weight five to six times. This rapid growth requires proper types of nutrition. The most common chick feed is called a chick starter and can be purchased from any local feed store or ordered online. The chick starter feed is usually 20% protein and filled with the necessary amino acids, vitamins, and minerals but doesn't include grit. The chick starter feed is ground up so that the chicks don't need additional grit to help digest it. Your chicks should remain on starter feed for the first 8 weeks, until they are introduced to grower feed.

The chick starter feed can come with medication to prevent coccidiosis. Some farmers with start the chicks on the medicated version for the first bag of feed and then switch to the non-medicated version. While feeding the non-medicated feed, it is often suggested to add apple vinegar to the chick water as a boost to the immune system and to serve as an antibacterial additive.

Waterers

Self-filling chick waterers are small with a saucer-shaped base and a screw-on plastic water jar. The waterer should be positioned to allow free access on all sides and placed on the opposite end of the brooder from the feeder.

Because the chicks can hop up and into the saucer, it is a common practice to put clean gravel or rocks in the trough to prevent accidental drowning during the first week or two.

Because the water can become contaminated with feed, shavings, and poop, the waterer should be replenished with fresh water each day. You can also place the waterer on top of a ½" board to help prevent the water from getting dirty as readily.

Some farmers choose to medicate the water by adding apple vinegar to the water twice per week, at a rate of 1 tablespoon per 1 gallon of water.

Brooder Duration

The chicks should remain in the brooder until they are fully feathered, which usually happens about week 6. This may need to be adjusted if the outside weather is less than 50 degrees F.

Flock Feeding

Feeding your flock is relatively easy with free choice, self-feeding equipment, but choosing the right type of feed and knowing when to switch to different rations are important.

Equipment

Chicken feeders are similar to chick feeders—just larger. One common feeder type has a large saucer at the bottom with a screw-on plastic feed container. The other common type is the long trough with a triangular top with large holes in the top for easy access.

Unfortunately, these types introduce several problems: waste from spillage, spoilage from wet weather, and rodent attraction. To alleviate these problems, many creative feeder alternatives are available online. For instance, some can be elevated, and others can be hung with an umbrella-type top to keep out the rain. The key is to have a viable source of feed readily available to the chickens at all times during the day.

Commercial Feeds

The most common feeds for chickens include commercial grower, layer, and broiler feeds. These feeds are especially formulated to meet the nutritional needs of chickens for egg and meat production.

Grower Feed

When the chicks reach an age of 8 weeks, they should be transitioned from the starter feed to grower feed. The grower feed is about 18% protein and designed to support continued growth and strong bone development until the birds reach maturity. The grower feed only contains 1.2% calcium. Growing chickens should not be fed layer feed, which contains a higher level of calcium. Too much calcium too early may result in kidney damage.

Layer Feed

Chickens generally are taken off the grower ration and started on the layer ration at about 18 weeks of age. This ration has 16% protein and is designed to support hens that are a laying age. The feed is formulated to meet the nutritional needs of laying hens, which includes an increase in calcium (2.5–3.5%) for strong egg production.

Meat-Bird Feed

Feed for meat birds can be fed to chicks from the first week and continued until harvested at optimum ages for broilers. The broiler feed has 22% protein to support rapid growth and meat production and achieve an optimum meat conversion ratio.

Supplemental Grains

If you want to supplement your commercial feed with whole grains to reduce cost, you can feed whole wheat, barley and/ or oats. The use of supplemental grains should not be more than 25% of the total diet, with formulated feeds making up the other 75%.

Limestone or oyster shells may also need to be included to ensure the laying hens are receiving enough calcium. Grit may also need to be supplied so that the chickens can properly grind the grains in their gizzard.

Digestive Process

The chicken initially stores grain in its craw or crop. Later, the food moves down the digestive tract to the stomach called the proventriculus. After the grain has been softened by digestive fluids in the stomach, the grain travels to the gizzard to be ground up with small pebbles. From there, it moves to the small intestines for the nutrients to be absorbed by the body.

Scratch Feed

As a fun addition to your formulated feed, scratch grains can be purchased to encourage natural pecking, foraging, and feeding instincts. It can be fed to laying hens or a mixed flock. It generally consists of whole grains, sunflower seeds, legumes, millet, and other natural ingredients. The total protein content is roughly 8%.

Because scratch feed is lower in nutrition, it should be fed in the afternoon after the chickens have had their nutritional needs met with a complete ration feed. Then only provide enough scratch grain that can be finished in 15–20 minutes.

Table Scraps

It is OK to feed chickens table scraps, such as peelings, over-ripe fruits and vegetables, watermelon rinds, etc. Like scratch grains, table scraps are low in nutrition and should not be fed more than what can be finished in 20 minutes. Excessive use of table scraps can adversely affect egg production.

Clippings

Pasture and lawn clippings are another source of supplemental snacks for your flock. Chickens are not able to digest fibrous plants but enjoy scratching through the clippings looking for seeds and bugs.

Flock Watering

Water consumption is one of the most important factors in chicken performance and health. The things that affect water consumption the most are water quality and water availability. To optimize water quality and availability, you should consider your water management and water equipment.

Water Management

Water consumption is important for digestion, waste elimination, and body temperature. Optimum water intake will also help to optimize feed intake, which is necessary for maximum growth and egg production.

Water quality will affect water consumption. Some of the things that can affect water quality are water pH, turbidity, mineral concentrations, bacteria, etc.

Water pH should be in the range of 5–6.8. Turbidity is the amount of foreign particle suspension and can affect taste. Minerals like sulfates and magnesium can be the cause of diarrhea. Bacteria can cause a host of health issues.

To accurately assess the quality of your water, you can have a water test done. Simple water test kits can be purchased online.

Water temperature and water cleanliness is also important and is something that can be managed with proper water equipment and maintenance.

Water Equipment

Cool, clean water should be freely accessible to your chickens at all times. Water helps to control body temperature and is important during hot weather.

Daily water intake volume will increase with age, so the size and number of waterers available will be an important consideration. On average, a mature hen will drink about 1 pint of water per day and up to 2 pints when it is hot. A 2-gallon waterer will service 8 chickens for a day. It is, therefore, recommended to have at least one waterer, on average, for every 8 chickens. If possible, waterers should be placed in the shade.

The most common types of waterers for chickens are the gravity flow, self-filling type; the automatic type; the water cup type; and the nipple type. When making your waterer choices, you should consider which option best performs the following.

- Provides cool water.
- Promotes clean water.
- Prevents algae formation.
- Allows for heating to prevent freezing.
- Allows for medication to be added.

Gravity Flow Waterers—Galvanized

The galvanized gravity flow waterers consist of a container and a corresponding round trough attached to the bottom of the container. The inner tank is filled with water from the top, and an outer shell fits over the inner tank and is locked into place with a turning motion. Water is then allowed to fill the trough, and the trough water level is vacuum controlled.

Most galvanized gravity flow waterers are designed to sit on the ground, in lieu of being hung. Because the water can become dirty if placed directly on the ground, it is recommended to elevate the trough level to about 5–6 inches by placing the waterer on a small platform.

Being double walled, the galvanized waterer is able to keep the water relatively cool in the summer and free of algae. For wintertime freeze protection, the galvanized waterer can be placed on an electric heated base that is thermostatically controlled.

Pros	Cons
Self-replenishing	Expensive
Durable, long-lasting	Water can get dirty
Keeps water cool	Requires refilling
Prevents algae	
Has heated base option	
2–5 gallons options	
Medication can be added	

Gravity Flow Waterers—Plastic

The plastic gravity flow waterers also consist of a container and a corresponding round trough attached to the bottom of the container. The round trough is screwed onto the container or attached by a turning lock. Trough water is vacuum controlled in some waterers and float controlled in other waterers.

Some plastic versions are filled with water by first flipping the container upside down and unscrewing or unlocking the trough. After the container is filled with water, the trough base is screwed or relocked into position and then flipped right side up. Other plastic versions have a screw top that will allow filling from the top without the flipping.

Some gravity flow waterers are designed to sit on the ground, and some can be suspended by a hanging handle. Because trough water can become dirty if close to the ground, it is recommended to either elevate the trough level to about 6–8 inches off the ground by placing it on a platform or by hanging the waterer to this height off the ground.

The water level is visible but can at the same time allow water temperature to be affected by the sun in the summer and allow algae to grow. For wintertime freeze protection, some plastic versions are electrically heated.

Pros	Cons
Self-replenishing	Some are difficult to fill
Relatively inexpensive	Sun can affect water temps
Can see water level	Sunlight can allow algae
Some can be hung	Not durable or long-lasting
Some can be heated	Water can get dirty
Some offer top filling	Some can freeze
Medication can be added	

Automatic Waterers

True automatic waterers are those that are connected to a water source and therefore do not require refilling a container or reservoir. They can come in a variety of designs, such as a trough, water cup, or nipple. Most trough-type delivery systems have a float valve for water level control. The cup type has a small flap, and the nipple types have a small toggle that controls water on demand.

Pros	Cons
Mostly maintenance free	Can't add medication
May allow for cleaner water	

Cup Waterers

This type of waterer option can be purchased with a ready-to-use container, or the cups may be purchased individually and added to a container of your choice. The cup has a small float flap, which when pushed by the chicken's beak will allow water to flow into the cup. This option is great for providing clean water on demand.

Pros	Cons
Relatively inexpensive	Limited usage at any one time
Allows for clean water	Can freeze
Reduced maintenance	

Nipple Waterers

The nipple waterer generally comes as a medium-to-large container equipped with one or more nipples on each side of the container. The nipple is a small toggle, which when pressed in or sideways allows a small flow of water on demand. This system may require a certain amount of learning time for your chickens to fully adapt.

Pros	Cons
Relatively inexpensive	Requires learning curve
Allows for clean water	Limited usage at any one time
Reduced maintenance	Can freeze

Egg Production

Chickens will begin to lay eggs about 20–24 weeks of age. Initial egg size will tend to be undersized but will become normal size after a few weeks. Chickens can lay eggs year-round without the need of a rooster.

The best laying years for a chicken are years 1–3. Depending on the breed, it is possible for a chicken to lay about 4–6 eggs per week, but numbers and frequency will slowly diminish with age. The breed will also affect the weekly number of eggs. Chickens can lay eggs for upward of 5–10 years, but if you are raising eggs to sell, you may want to arrange a replacement schedule for your hens to be replaced every 2–3 years.

Chickens generally lay their eggs in the morning within the first 6 hours following sunrise. The best egg production scenario includes the use of a coop with nesting boxes.

Egg Production Enhancement

Optimum egg production of your homestead will require the consideration and implementation of key factors, such as breed selection, housing and space, nesting boxes, egg collection, nutrition, free-range limitation, lighting, cleanliness, health, and pest and predator control.

Best Breeds

Some breeds are best for egg production, some are best for meat, and some are used for dual purposes. Your breed choice will depend on your preferences and goals.

Housing and Space

A clean and well-ventilated coop will be needed with adequate space for each chicken. Each bird should have at least 3–4 square feet of space in the coop and 8–10 square feet of space in the outdoor run.

Nesting Boxes

You should have a chicken coop with nesting boxes for the hens to lay their eggs. You should have one nesting box for every 3–4 hens. These boxes should be dark, quiet, and comfortable to encourage egg laying. The best nesting material in the boxes is wood chips or sawdust.

Egg Collection

Ideally, eggs should be collected regularly, at least once or twice a day, to prevent them from getting dirty or damaged. Eggs should then be cleaned and promptly stored in the refrigerator.

Proper Nutrition

To maximize egg production, a balanced, high-quality chicken feed that meets the nutritional needs of your chickens should be available as free access. A layer feed with at least 16% protein and 2.5–3.5% calcium is essential for good egg production. Fresh, clean, water available at all times is also essential.

Free-Range

Some homesteaders choose to allow their chickens access to outdoor areas where they can forage for insects, grass, weed seed, and other natural foods. This not only helps to supplement their diet but also helps to keep them healthy and happy. For better egg accountability, some choose to allow free-ranging for a short time after eggs have been laid in the egg boxes.

Lighting

Chickens require a certain amount of daylight to trigger egg production. Ideally, the chickens need 14–16 hours of daylight in order to maintain a proper egg laying schedule. When natural daylight is insufficient, artificial lighting can be added in the coop to achieve the needed daylight hours.

Cleanliness

To keep your chickens healthy and productive, you should maintain a regular routine of keeping the coop, egg boxes, and chicken run clean of poop and soiled organic material. You should also ensure that the water supply is clean and accessible.

Health

You should maintain a vaccination and healthcare schedule for your chickens, and consult a veterinarian when health issues arise.

Pest and Predator Control

You should protect your flock from pests and predators by elevating the coop off the ground and securing the coop and the run with strong fencing. Chicken feed is a strong attractant to rodents. Larger rodents can eventually be attracted to the eggs as well. Infestations can be avoided with proper vigilance and the use of proper control measures.

Low Wintertime Egg Production

Chickens often lay fewer eggs in the winter, and, in some cases, they may even stop laying altogether. There are several reasons for this seasonal decline in egg production.

Reduced Daylight

Chickens are influenced by the amount of daylight they receive. As the days get shorter in the fall and winter, hens receive fewer hours of natural light, which can signal their bodies to slow or stop egg production.

Temperature and Weather

Cold temperatures can stress chickens and cause them to direct their energy toward staying warm instead of egg production. Extremely cold weather can also lead to frozen water sources, making it difficult for chickens to stay properly hydrated, which can impact egg production.

Molting

Many chickens go through a molt in the late summer or fall. During this time, they shed their old feathers and grow new ones. Molting requires a lot of energy, and hens often pause egg laying during this period.

Seasonal Changes

It's worth noting that not all chickens will completely stop laying eggs in the winter. Some breeds and individual hens may continue to lay at a reduced rate. The probability can be enhanced if they are kept in a well-managed environment with supplemental lighting, a suitable diet, and appropriate shelter from the cold.

To encourage winter egg production, some homesteaders use artificial lighting to extend the daily light period and ensure their chickens have access to a nutritious diet. However, it's important to not push your chickens too hard—they do benefit from a natural break in egg production during the winter months.

Chicken Coop

Chicken coops are structures that provide an enclosed area to roost at night, lay eggs, and be protected from harsh weather and predators. The coop should include nesting boxes for laying eggs, windows for lighting and ventilation, a floor covered with litter material, perches for roosting, and lighting for controlling daylight hours.

Space Requirements

On average, chickens need about 3–4 square feet per bird inside the coop. The floor should be covered with a form of litter that is absorbent. The litter will help to keep the floors clean and reduce odor.

Nesting Boxes

Nesting boxes are used by the chickens to lay eggs. The size should be roughly 12 inches square. To facilitate egg collection, the boxes are normally constructed side by side and are usually built 18–20 inches above the floor into and through the back wall with an outside hinged top or end for easy access.

You should plan on having one box for every 3 hens. The box should be lined with either wood shavings or sawdust to encourage nesting and laying. The litter also helps to protect the eggs from being damaged. If boxes are being used more than once in the morning, you may want to collect eggs twice a day to prevent the danger of breakage.

Bedding

The best bedding material for the coop floor is wood shavings about 2–3 inches deep. The bedding should be replaced every 2–4 weeks.

Roosting

Typically, chickens like to roost on something elevated above the ground for sleeping at night. The best perches for roosting are 2" x 4"s that are arranged horizontally to the floor inside the coop.

An easy design is to cut two, 2" x 4"s, 4 ft. long, which will be used to support the perches. Determine where you want your roosting bars or perches to be situated, and place the 4 ft. supports on each end, about 4 feet apart. Arrange for the top of the support to be attached to the wall, about 42 inches above the floor, and then angle out at roughly 45 degrees, so that the bottom of the support can be attached to the floor, roughly 24 inches from the wall.

Next, cut two notches in each support (like a staircase), on which the two perches are to be placed. The bottom perch should be placed roughly 16 inches above the floor and 16 inches from the wall. The upper perch can be situated roughly 30 inches above the floor and 8 inches from the wall. Cut two perches, 6 feet long, and place them into the notches in the supports; then attach them to the two end supports. You will have two perches, 6 feet long, that will extend 1 foot beyond the supports on each end. This roosting arrangement will allow each hen to have 12 inches of roosting space.

The hens will poop the most while roosting, and, therefore, a lot of poop can build up under the roosting perches. A latched, trap door can be built into the floor under the perches. Such an arrangement allows for easy access when the floor needs to be cleaned. Just open the trap door and allow the soiled litter to fall to the ground for easy removal. Then relatch the door closed and add fresh litter inside.

Construction

There are many ways to construct a chicken coop, and you can find a variety of plans online. You can also find many prefabricated kit versions online. The most common plans are those that use wood for construction material. As mentioned earlier, the space requirement for the coop is 3–4 square feet per bird. For this example, a coop designed to house 12 chickens will be used. This would translate to a 48-sq. ft. coop.

Site

The first step in building a coop is site selection. It is best to try to find a site that is as level as possible. Even though you wouldn't want a coop adjacent to your home, you might want to have it relatively close for ease for egg collection in the winter and to deter predators.

Floor

The best plans for a coop are those that are elevated off the ground. The advantage of having the floor elevated is to reduce the threat of rodents and snakes, reduce moisture problems when it rains, and facilitate soiled litter removal with a trap door.

To be elevated, it is best to begin by installing 6-inch diameter round posts or 4' x 4' treated timber posts in the ground at all four corners. The posts should be secured in the ground with concrete and should extend about 14–16 inches above the ground. To accommodate a 48-sq. ft. structure, the corner posts should be 6 ft. apart on the sides and 8 ft. apart on the front and back.

The frame for the floor would be constructed with 2" x 6" boards and rest on top of the four corner posts. The frame would consist of the outer frame and inner floor joists. Then 4' x 8' sheets of ¾" plywood can be cut and applied to the top of the frame for the floor.

Walls

For this example, the front wall will be 6 ft. tall and the back wall 5 ft. tall. The walls can be framed with 2" x 4"s, and 4' x 8' sheets of ½" plywood can be cut and applied to the outer frame for the walls. Windows can be cut through the walls for ventilation. The windows are typically above the roost in the front and above the nesting boxes in the back for cross ventilation. The windows should be covered with screen or hardware cloth to keep out rodents and predators and should have shutters that can be closed at night and during harsh weather.

Roof

The roof can be arranged to be gabled in the front and back with the pitch slanting toward each side, or you can have the front wall higher than the back wall so that the roof slants toward the back. There are a variety of methods and models to choose from. For this example, simply arrange for the front wall to be about 1 ft. higher than the back wall, allowing the roof to slant toward the back with a 2–12 pitch.

The nesting boxes will be extending through the back wall for easy access, so the roof can either be extended outward an extra foot, so that rainwater doesn't drip on the top of the nesting boxes, or you could add a gutter at the back.

Doors

There are many options to choose from, but most prefer to have a small door at the front corner with a ramp to the chicken run for the chickens to have easy access in and out. In addition, a separate, larger door would allow access for the homesteader to enter for cleaning and maintenance.

The doors should have latches, which can be closed during the night. The latches should be secure and lockable to keep out clever animals, like raccoons.

Ramp

If the coop is elevated, you will need to build a ramp, so that the chickens can have easy access in and out. The ramp can be made of wood. The ramp should be built in such a way so that it doesn't sag in the middle and should have 1" x 2" cross boards on the surface to prevent slipping when wet.

Electricity

The coop should be wired for electricity, so that you can add lighting during the winter when daylight hours are less than the needed 14 hours. It is also handy to be able to add electric heaters when cold weather gets below what is acceptable for your flock.

Chickens can keep themselves warm with their feathers but will still need supplemental heat when temperatures get down in the 30s (Fahrenheit). The chickens might survive freezing weather, but the eggs will not.

Feed and Water

There is a lot of debate about whether feeders and waterers should be included in the chicken coop. The prevailing opinion is that the feeders and waterers should be in the chicken run and not in the coop.

It really depends on the weather. During harsh or wet weather, you might need to have feed and/or water available in the coop—at least until the weather improves and the chickens are able to get out into the run.

Chicken Run

Every chicken coop should have a chicken run attached. The chickens need to be outside during the day, and an enclosed chicken run allows the chickens to be contained for feed and water management and protected from predators.

On average, the run should be big enough to provide at least 10 square feet per chicken. Of course, it is OK to have more than that. For this example, the chicken run is for 12 chickens and is 16 ft. x 8 ft. (128 sq. ft.).

Enclosure Construction

The typical run consists of wooden posts, hardware cloth, welded wire fencing, and chicken wire. The posts can be 6-inch diameter round 8 ft. posts or 4" x 4" x 8' posts. The posts should be installed 2 ft. in the ground with 6 ft. aboveground. The posts should be installed 6–8 ft. apart.

A 1-ft. trench should be dug along the outside perimeter and horizontal to the run. This will allow hardware cloth to be extended around the perimeter of the run, from post to post, and extend down into the trench. The hardware cloth fencing can be stapled to the post with fencing staples. You should end up with hardware cloth fencing material 1 foot into the ground, and 2 ft. aboveground, around the perimeter of the run.

You can then refill the trench with dirt and tamp it in. Once done, you can continue with fencing the perimeter walls of the run with welded wire fencing. This heavier gauge of wire is much stronger than regular chicken wire and will provide better long-term protection from predators.

Wooden 2" x 4" x 8' boards should be placed on top of the posts, running horizontal to the run to form a top plate. Wooden 1" x 4" x 8' boards can then be placed on top of the posts, from side-to-side, to provide purlins running perpendicular to the run. This top frame structure will allow for adding chicken wire on top of the run.

It is always a good idea to have the first 4–8 feet of the run covered if possible. This provides shade from the sun and can prevent your feed from getting wet when it rains or snows.

With that in mind, you can add two more 1" x 4"s to the top—one next to the coop, and the other halfway between the coop and the first set of posts. You can then add two sheets of 4' x 8' plywood and secure them to the 1" x 4" purlins running perpendicular and the 2" x 4" top plate running horizontal to the run.

For the remainder of the run, you can lay a 4-ft. roll of chicken wire on top of the run. You can roll out the chicken wire twice horizontally (side-by-side) and attach the two together with hog rings, and then staple the chicken wire to the purlins and top plate with fencing staples. For added protection, you can add a second layer of chicken wire on top of the first layer, but lay the second layer perpendicular to the first layer. Again, connect the two strings of wire together with hog rings, and staple the wire to the purlins and top plate. That way, you have two layers of chicken wire on top of the run, perpendicular to each other, for added protection from raptors and owls.

This type of chicken run enclosure can be self-fabricated for roughly $250 or about $2 per sq. ft. This, of course, depends on prices for material found locally.

Prefabricated Runs

You can purchase prefabricated chicken runs that are made with galvanized pipe and covered with a chicken wire-type mesh with additional shade cloth on one end. They come in a variety of shapes and sizes. A 200-sq.-ft. enclosure can be purchased for about $250.

Unique Behavior

Poultry can demonstrate unique behavior from time to time. Some of these should be discouraged, while others should be accommodated.

Pecking

It is not uncommon to have dominant hens that will try to establish a pecking order, and in doing so will pick on smaller hens. Some common reasons for pecking are hierarchy, stress, overcrowding, and not enough protein in the diet.

Stress can be caused by summer heat, overcrowding, or boredom during winter months when being cooped up too long. You should ensure that the coop has proper ventilation, and you should verify you have at least 4 sq. ft. of space per hen in the coop, and 10 sq. ft. per hen in the run.

You should also check their diet to ensure that optimum protein percentages are being met. Parasites can be another reason for pecking other birds or themselves. You should regularly inspect your birds for parasites or illness.

If one particular hen is doing all of the pecking, or if one particular hen is being pecked by several hens, the aggressive hen and/or the victimized hen should be isolated and may need to be removed from the flock.

Broodiness

A hen will exhibit broody behavior by sitting on a clutch of eggs to hatch them, even when the eggs are not fertilized. Though this behavior is a natural instinct, it can become problematic if you are wanting to collect your eggs regularly.

There are several things you can do to dissuade this behavior. One of the simplest ways to discourage brooding behavior is to remove eggs from the nesting boxes as soon as they are laid. Hens may become less inclined to sit on empty nests.

As a deterrent, you can make the nesting boxes less appealing by changing their layout or making them less comfortable by adding straw instead of sawdust. Hens prefer to brood in a dark, secluded spot, so another effective deterrent is to make the nesting box less dark by adding an artificial light.

If the broody behavior persists, you may try replacing the natural clutch eggs with imitation eggs, which can eventually signal the hen that the eggs aren't viable and, therefore, end her desire to brood.

As a final solution, you may have to remove the hen from the coop in the morning and make her stay outside during daylight hours. The key is to be patient—the broodiness will pass in time.

Dust Bathing

Dusting is a behavior commonly exhibited by chickens, where they bathe themselves in dust or loose soil. Chickens will often find a dry and dusty area in the ground and then proceed to flap their wings, roll around, and create dust clouds. This behavior serves several purposes. The most common are to help to control parasites and pests, such as mites and lice; to help regulate body temperature; and to reduce stress.

Overall, it is a good practice to provide a place where the chickens can enjoy this ritual. One common method is to fabricate a dust box, much like a sandbox for children. A 2 ft. square box can be fabricated with 2" x 6" boards and then filled with a fine, silty soil. Silt, by itself, is dusty. To keep it viable, you will need to keep the area clean and provide a top for protection when it rains.

Roosters

Hens do not need a rooster to lay eggs, but they do need a rooster to fertilize eggs. If you have a large flock and plan on replacing old hens each year, or if you plan on selling chickens regularly as a form of revenue and want to incubate your own eggs, then you will need a rooster for fertilization.

However, if you have a small flock, like the example flock of twelve, it might be more cost-effective to replace your old hens from time to time by purchasing chicks from a reliable hatchery.

There are a number of advantages and disadvantages of having a rooster included in your flocks. For homesteads that have small flocks, the disadvantages often outweigh the advantages. For homesteads with larger flocks, however, the opposite can be true.

The most common scenarios where roosters are welcomed are homesteads that are able to add free-ranging as part of their flocks' daily routine. For those that have roosters for fertilization purposes, you should have 1 rooster for every 10 hens or so.

Pros	Cons
Rooster can fertilize eggs.	Roosters are loud and can disturb neighbors.
Roosters are known to protect the flock from predators.	Roosters can be aggressive toward people.
	Roosters can be too assertive toward hens.

Healthcare

Common health issues in homestead chickens can vary, but some of the most prevalent ones include the following.

Respiratory Infections

Chickens can suffer from respiratory illnesses, such as respiratory viruses and *mycoplasma* infections. These can be prevented by maintaining good ventilation in the coop. Treatment may involve antibiotics prescribed by a veterinarian.

Parasitic Problems

External parasites, like lice and mites, as well as internal parasites, like worms, can affect chickens. Regular cleaning of the coop, providing clean bedding, and administering dewormers as necessary can help prevent and treat these infestations.

Having access to a dust bath is also helpful in deterring external parasites in your flock. You can build a small 2' x 2' dust box out of 2" x 6" boards. The box should be filled with loose, silty soil, to allow for the flock to bathe in the dust that they create for themselves. A removable top is recommended to prevent the soil from turning to mud when it rains.

Coccidiosis

This is a common intestinal disease in chickens caused by protozoa. Good sanitation and the use of medicated feed or water can help prevent coccidiosis. Treatment may involve medication prescribed by a vet.

Egg-Laying Issues

Problems like egg binding or soft-shell eggs can occur when your hens don't have enough calcium. For severe cases of egg binding, you may need to consult a veterinarian.

Heat Stress

Overheating can be a significant issue, especially in hot climates. Providing shade, cool water, and good ventilation can help chickens cope with heat stress.

Injuries

Chickens can sustain injuries from being pecked by other hens, predator attacks, or accidental mishaps. In such cases, you should isolate the injured hens, if necessary; clean and disinfect wounds; and consider consulting a veterinarian for severe injuries.

Diseases Like Avian Influenza or Newcastle Disease

These contagious diseases can have serious consequences for flocks. Some of the symptoms for avian flu include lethargy, lack of appetite, and/or diarrhea, which might result in sudden death. Symptoms for Newcastle disease include sneezing, coughing, droopy wings, and less-to-no egg production.

These diseases are contagious and thus considered a health risk to neighboring flocks in the local area as well as your own. If you suspect 1 or more hens are demonstrating symptoms, you should isolate the sick birds and follow any vaccination guidelines provided by local agricultural authorities.

If you have a large flock, it's a good idea to have a relationship with a local veterinarian who is knowledgeable about poultry health to help with diagnosis and treatment when necessary.

Expenses for Eggs

The following expense data will be based on an example flock of 12 hens, beginning by purchasing a dozen chicks from a hatchery to be used as laying hens.

Start-Up Expense Summary Charts

Start-up expenses will include the initial investment in chicks and the cost for equipment and housing.

Initial Poultry Investment Expense Chart

Item	Cost
12 chicks @ $2.50 per chick	$30

Initial Feed Investment to Get Pullets to Laying Age

Starter Feed

Chicks will be fed chick starter crumbles at 20% protein for 8 weeks. On average, each chick will consume roughly 1 oz. per day for 56 days.

- 1 oz. per day for 56 days = 56 oz.
- 56 oz. ÷ 16 oz./lb. = 3.5 lb. total per chick for 8 weeks
- Average price of $25 for 50 lb. = $0.50 per lb.
- 3.5 lb. @ $0.50 per lb. = $1.75 per chick for first 8 weeks.

Grower Feed

Pullets (young hens) will be fed grower feed for 10 weeks (weeks 9–18). For this 10-week period, the pullets will consume an average of 3 oz. per day of grower feed.

- 3 oz. per day for 70 days = 210 oz.
- 210 oz. ÷ 16 oz./lb. = 13 lb. total for 10 weeks
- Average price of $28 for 50 lb. = $0.56 per lb.
- 13 lb. @ $0.56 per lb. = $7.28 per pullet for 10 weeks

Item	Cost per Bird	Cost for 12
Feed cost for weeks 1–8	$1.75 per chick	$21.00
Feed cost for weeks 9–18	$7.28 per pullet	$87.36
Total	**$9.03**	**$108.36**

Infrastructure and Equipment Expense Chart

Item	Cost
Brooder box	$45
Heat lamp for light	$18
Brooder heat plate	$45
Chick feeder & waterer set	$18
Hen feeders	$25
Hen waterers	$35
Chicken coop—($13.50/sq. ft.)	$650
Chicken run—($2.00/sq. ft.)	$250
Dust bath	$15
Total	**$ 1101**

Operating Expenses

Operating expenses will include layer feed for a small flock of 12 laying hens. This example will not include a rooster.

Feed Costs

Laying hens will eat, on average, about ¼ lb. of feed per day. Layer feed can be purchased for $16 for a 40-lb. bag. The following chart will be feed cost for 12 hens.

- .25 lb. per hen for 12 hens = 3 lb. feed per day
- 3 lb. for 7 days = 21 lb. for 12 hens per week

- Average price of $16 for 40 lb. = $0.40 per lb.
- 21 lb. @ $0.40 per lb. = $8.40 for 12 hens per week

Item	Daily Cost	Weekly Cost	Annual Cost
Layer feed for 12 hens	$1.20	$8.40	$436.80

Animal Care

Bedding

Wood shavings can come in bales of 8 cu. ft. Each bale costs roughly $8. One bale is needed to cover the floor of a 48 sq. ft. coop, 2 inches deep. The bedding should be replaced every 2–4 weeks. If the bedding is replaced every 2 weeks, the weekly cost would be $4. Sawdust for the nesting boxes will cost roughly the same but will last 4–6 times as long.

Vet Care

Vet care can vary, but the prevailing cost estimate is the $52 per year range for a dozen hens.

Item	Weekly Cost
Nesting box bedding (sawdust)	$1.00
Coop bedding (wood shavings)	$4.00
Vet care	$1.00
Total	**$6.00**

Cost/Benefit for Eggs

The purpose for this analysis is to provide a way to examine the cost versus the benefit of raising chickens for household eggs.

The analysis will compare the weekly cost for egg production to the cost of purchasing eggs from a store.

Savings/Loss

Even though it is possible to get roughly 6 eggs per week per hen, the reality is closer to 4 eggs per week per hen. With that in mind, it can require 3 hens to produce a dozen eggs per week. For a flock of 12 hens, that equates to 4 dozen eggs per week.

A family of four could eat 2 dozen per week and sell 2 dozen per week. The value of 4 dozen eggs per week, based on local prices, would be roughly $3.50 per dozen.

- **Note:** The price per dozen is an estimate and can vary depending on the locale and the economy.
- **Note:** Operating costs only are used in this analysis—start-up costs are not included.

Item	Weekly Expense	Weekly Value or Revenue	Weekly Savings/ Loss
Feed & misc.	$8.40		
Animal care	$6.00		
Total cost for 4 doz. eggs	**$14.40**		
Egg value per 4 doz. eggs		$14.00	
Total value/revenue		**$14.00**	
Savings/loss			- $.40

As you can see, the cost/benefit for raising your own eggs is a break-even proposition, and this isn't taking into consideration the start-up cost and value of labor. Based on the data above, the cost per dozen is about $3.60, which is about $0.10 more per dozen than purchasing the eggs at the store.

Nevertheless, the major benefit of raising chickens for eggs is twofold. The quality of the eggs is probably better, and you also have the enjoyment and the peace of mind of being self-sustaining and raising your own food off the grid.

Expenses for Meat

The following expense data is based on raising broilers for meat, beginning by purchasing chicks from a hatchery.

Start-up Expenses for Meat

Start-up expenses include hatchery chicks and feed. Formulated feed for meat birds will be fed free-access to the chicks from week 1 until they leave the brooder.

Operating Expenses for Meat

Meat chickens will be fed "free access" meat bird feed from the brooder until harvest age. Animal care will be minimal for short term broilers.

Cost/Benefit for Meat

The purpose of this analysis is to provide a way to examine the cost versus the benefit of raising chickens for household meat. The analysis will use data for the Cornish Cross, and compare the 8-week cost for production to the cost of purchasing meat retail.

Savings/Loss

Raising chickens for meat is a common occurrence on home-steads. Meat chickens are usually harvested as broilers. After leaving the brooder at 4–6 weeks, they are fed additional weeks for finishing until arriving at the optimum weight.

This optimum weight and harvest age depends on the breed. For demonstration purposes, the following data set is for the Cornish Cross breed, which is often raised for meat due to its exceptional growth rate and performance. The Cornish Cross is typically harvested at 6–8 weeks with a dressing weight of 4–6 lb.

Meat chickens consume much more feed than layer hens. The feed for meat chickens is a high-protein feed specifically formulated for fast growth. This allows breeds like the Cornish Cross to have a good feed conversion ratio of 2:1. That means a Cornish Cross broiler chicken yielding 6 lb. of meat will have consumed roughly 12 lb. of feed.

Meat-bird feed, with 22% protein, costs roughly $20 for a 50-lb. bag. This would be $0.40 per lb. The total cost for feed per Cornish Cross broiler is $4.80 (12 lb. x $0.40 = $4.80).

With this information, you can determine the cost per lb. for each broiler produced and the savings/loss per broiler produced.

Item	8-Week Expense	8-Week Yield in Meat	Retail	Savings/ Loss
Feed & misc.	$4.80			
Cost per chick	$2.50			
Total cost per broiler	**$7.30**			
Meat at 8 weeks		6 lb.		
Cost per lb. meat processed	**$1.22**			
Cost per lb. retail			**$3.49**	
Savings/loss				**+ $2.27**

Please note that the total cost per broiler does not include the cost for processing. Paid processing is roughly $3.50 per bird, which would raise the cost per lb. to $1.80 and reduce savings benefit to $1.69.

A detailed description of how to process your own meat birds is found in the Meat Processing section (for Geese) in Chapter 3, Part 2.

The above data set is for the Cornish Cross. Other breeds take longer to arrive at their optimum weight and age. The longer the production time, the higher the feed conversion ratio will be.

Picture Gallery

Egg Chickens

Rhode Island Red

Leghorn

Sussex

Plymouth Rock

Australorp

New Hampshire

Wyandotte

Meat Chickens

Cornish Cross

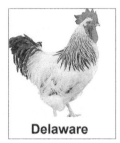
Delaware

Freedom Ranger

Jersey Giant

Orpington

CHAPTER 2
Ducks

Ducks are often considered a viable option for eggs on a homestead, because ducks can actually lay more eggs than chickens, but ducks do come with a set of caveats.

If space is limited, ducks may not be the best option because their poop is so messy. However, if foraging is a possibility, then raising ducks for eggs and/or meat can be a feasible option.

In this chapter, the most common duck breeds used for eggs and meat are compared. Feeding, watering, and egg production, as well as shelter and containment requirements are discussed, and a list of the common health issues and how to avoid them is reviewed.

As a conclusion, a cost/benefit analysis is provided in order to assist you in the decision-making process. Some cost/benefit adjustments are suggested that can render your duck raising more feasible.

Breeds

There are many varieties of ducks to choose from in the US for homesteads. The most common breeds used for eggs, meat, and dual purpose are listed below. The descriptions and charts will help you decide which breed best meets your needs.

Breeds for Eggs

These duck breeds are popular choices for homesteaders due to their valuable traits, including egg production, adaptability, and friendly temperament.

Khaki Campbell

The Khaki Campbell originated from England. The breed is a cross between the Rouen and the Indian Runner. They have khaki-colored plumage and are a medium-sized breed. Khaki Campbells are known for their excellent egg-laying abilities. They can lay about 300–340 white or khaki-colored eggs per year. Khaki Campbells are popular for their calm and friendly nature, making them suitable for homesteads.

Indian Runner

The Indian Runner originated from India. Indian Runner ducks are tall, slender ducks that stand upright. This breed comes in various colors, including white, fawn, and chocolate. Indian Runners are generally active and alert ducks. They are prolific layers, producing approximately 250–300 white eggs annually.

Welsh Harlequin

The Welsh Harlequin originated from Wales. They are a medium-sized duck and are known for their attractive and unique plumage of silver and gold coloration with distinct markings. Welsh Harlequins are known for being docile and good-natured. Welsh Harlequins are known for their high egg production with an average of 240–330 pale blue or greenish eggs per year.

Ancona

The Ancona originated from Italy. Ancona ducks are medium-sized and are known for their distinctive black and white mottled plumage, giving a speckled appearance. Anconas are hardy and adaptable, making them well-suited for homestead environments. They are prolific layers, producing about 200–280 white or tinted eggs per year.

Breeds for Meat

Both Pekin and Rouen ducks are popular choices for meat production due to their relatively large size and excellent meat quality. The choice between them may depend on personal preferences and regional availability.

Pekin

The Pekin duck is believed to have originated in China. Pekin ducks are medium-to-large-sized ducks. They have white plumage, orange bills, and orange legs. Adult Pekin ducks typically weigh between 8 and 11 pounds. Because Pekin ducks are known for their calm and friendly temperament, they are often kept as pets and are good foragers.

Pekin ducks are highly valued for their meat, which is known for being tender, succulent, and flavorful. They are a popular choice for homestead meat production.

Rouen

The Rouen duck is a breed developed in France and is named after the city of Rouen. Rouen ducks are large ducks, similar in size to the Pekin. They have plumage that resembles that of wild Mallard ducks, with feathers in shades of brown and green. Adult Rouen ducks typically weigh between 8 and 10 pounds. Rouen ducks are generally calm and docile, making them suitable for backyard and small farm settings.

Rouen ducks are a good choice for meat production. They are valued for their meat, even though their meat is often considered slightly gamier compared to the Pekin.

Dual-Purpose Breeds

These duck breeds vary in terms of size, plumage, temperament, egg production, and meat quality, so choosing the right breed depends on your specific needs and preferences.

Cayuga

The Cayuga duck originated in the United States, particularly in the Cayuga Lake region of New York. They are medium-sized ducks and have black plumage with iridescent green highlights. Adult Cayuga ducks typically weigh between 6 and 8 pounds. They are generally calm and friendly. Cayuga ducks can lay about 100–150 eggs per year. They are valued for their flavorful meat, which is dark and rich.

Silver Appleyard

The Silver Appleyard duck was developed in England. They are medium-to-large-sized ducks and have striking silver and white plumage. Adult Silver Appleyard ducks typically weigh between 7 and 8 pounds. They are usually friendly and good foragers. Silver Appleyard ducks can lay about 150–200 eggs per year. They are considered a dual-purpose breed, valued for both meat and eggs.

Saxony

The Saxony duck was developed in Germany. Saxony ducks are medium-sized. They have a distinct silver-blue plumage. Adult Saxony ducks typically weigh between 6 and 7 pounds. They are generally calm and docile. Saxony ducks can lay about 200–250 eggs per year. They are considered a dual-purpose breed with good meat and egg production.

Buff Orpington

Buff Orpington ducks are a variety of the Orpington breed, which originated in England. They are medium-sized ducks and have a vivid, solid buff color. Adult Buff Orpington ducks typically weigh between 7 and 8 pounds. They are generally friendly and calm. Buff Orpington ducks can lay about 150–200 eggs per year. They are a good choice for meat production due to their size and meat quality.

Magpies

The Magpie duck breed originated in Wales. Magpie ducks are medium-sized, and they have a distinctive black and white pied plumage. Adult Magpie ducks typically weigh between 4.5 and 6 pounds. They are usually active and hardy. Magpie ducks can lay about 180–220 eggs per year. They are considered a dual-purpose breed, suitable for both meat and egg production.

Black Swedish and Blue Swedish

The Swedish ducks, including Black and Blue varieties, are from Sweden. They are medium-sized ducks and are bluish-gray in color. Adult Swedish ducks typically weigh between 6 and 7 pounds. They are known for being hardy and adaptable. Swedish ducks can lay about 150–200 eggs per year. They are suitable for meat production, especially the Black Swedish, which is valued for its dark meat.

Breed Comparison for Eggs

Criteria	Khaki Campbell	Indian Runner	Welsh Harlequin	Ancona
Eggs per yr.	300–340	250–300	200–330	200–280
Eggs per wk.	5–6	4–6	4–6	4–5
Egg size, and color	Medium, Large khaki to green	Medium, white	Medium, blue - green	Medium, off white

Breed Comparison for Meat

Criteria	Pekin	Rouen
Weeks to processing	6–9	10–12
Processed wt. (lb.)	5–7	5–7

Dual-Purpose Breeds

Criteria	Cayuga	Silver Apple- yard	Saxony	Buff Orping- ton	Magpie	Black & Blue Swedish
Eggs per week	2–3	4–5	4–5	2–3	3–4	3–4
Egg size and color	Large dark	Medium white	Medium white	Large tan	Medium white	Medium white
Weeks to processing	9–12	8–12	8–12	8–12	8–12	8–12
Processed wt. (lb.)	4–6	4–6	4–6	4–6	4–6	4–6

Hatching and Brooding

The incubation and brooding process is similar to that of chickens. For a detailed description of the process and various equipment options, refer to Starting with Eggs and Starting with Chicks in Chapter 1, Part 2.

Incubating Eggs

Basically, the same incubator and method of use can be used for both chickens and ducks. The only difference for duck eggs is the humidity and the number of days required for hatching. Duck eggs require, on average, 28 days to hatch. Duck eggs should be incubated at a temperature of 99–99.5 degrees F. for 28 days.

Humidity should be 55–65% for the first 24 days. On day 25, egg turning should stop, and the humidity should be increased to 65–75% until day 28. When pipping starts, the humidity should be increased to 80–85%.

Brooders

The process and equipment for brooding ducklings are similar to that for chicks. The main differences are bedding and water. Ducklings will remain in the brooder for 6–8 weeks.

Brooder Space

The space requirement per duckling is about 1 sq. ft. per duckling for the first 3 weeks. For weeks 4–8, each duckling will require about 3 sq. ft.

Brooder Heat

The temperature should begin at 90 degrees F. and be reduced 5 degrees per subsequent week thereafter. Like for chicks, the heat plate works best as a heat source, as it allows for daylight hours management. The light management would be the same as for chicks.

Brooder Water

Because ducklings tend to be messy and splash the water, a special accommodation is required. Ducklings like to splash and dip their heads in the water. This creates quite a wet mess around the waterer area. To accommodate this, a good brooder choice can be the galvanized metal livestock water trough that is rectangle in shape. The size can be 2 ft. tall x 2 ft. wide x 4 or 6 ft. long.

For water management, you can also place a metal cookie baking pan with a ¾-inch lip on the opposite end of the brooder from the feeder. You can fill the pan with gravel, and then place the waterer on top of the gravel. This will help to contain the splashed water and assist in preventing the bedding from getting too wet. You will need to clean the pan and change the water more frequently than you would for chicks. The same type of waterer that is used for chicks will work fine.

Brooder Bedding

Duck poop contains a lot of moisture. For that reason, bedding can become moist and messy quite frequently. With that in mind, wood pellets can be a better alternative for bedding than wood shavings. The wood pellets will absorb the moisture much better and last longer than wood shavings.

The top of the brooder should be covered to protect the ducklings. Typically, the top would consist of a metal frame covered with hardware cloth or chicken wire. The most common predators for brooders are cats.

Brooder Feeding

During the first week, each duckling will consume about .15 pounds of feed. Between weeks 2–8, the feed consumption per duckling will gradually increase from .15 lb. to 2.5 lb. of feed per week. During their 6–8 week stay in the brooder, the ducklings should have continuous access to feed and water.

The same bulk type of feeder that are used for chicks will work for ducklings in the beginning. After about 3–4 weeks, you may need to switch to a metal trough type feeder, so that the ducklings can have full access to feed at all times.

As a feed choice, baby ducklings should be fed starter crumbles, which are specially formulated for their growth and development. Ducklings need more protein than adult ducks, especially in the first 2 weeks of their lives. The duckling starter feed has a protein content of 20–22%, which is ideal for the first 2 weeks. For weeks 3–8, you can reduce protein content to roughly 15–18%. Duckling grower feed can be used for weeks 3–8, in a small pellet form, with higher niacin content.

Flock Feeding

Feeding your ducks is relatively easy with free choice, self-feeding equipment, but it is important to use the best formulated feed at the correct stage of growth for optimum development.

Equipment

Duck feeders can be the same as chicken feeders. One common feeder type has a large saucer at the bottom with a screw-on plastic feed container. The other common type is the long trough for easy access. The trough type seems to be easier to keep clean.

Unfortunately, these types of feeders introduce several problems: waste from spillage, spoilage from wet weather, and cleanliness from poop. To alleviate spillage and to help keep the poop out of the feeders, one can elevate the feeders by placing them on bricks or thick boards.

To prevent spoilage from wet weather, you can install a metal or plastic cone-shaped cover over the top of the feeders to deflect rainwater away from the feeders. The key is to have a viable source of feed readily available to the ducks at all times during the day.

Commercial Feeds

The most common feeds for ducks include commercial starter, grower, developer, layer, and maintenance feeds.

Feeding ducks requires specific types of commercial feed that are formulated to support their growth and nutritional needs at different age stages.

Starter Feed (Weeks 1–2)

Newly hatched ducklings should be fed a high-protein starter feed containing about 20–22% protein. This helps promote early growth and feather development.

Note: It is not recommended to feed your ducklings medicated chick starter. Ducklings eat more than chicks and can overmedicate themselves. Chick feed is medicated to prevent coccidiosis, which does not typically affect ducklings.

Grower Feed (Weeks 3–7)

As the ducklings grow, you will need to transition them to a grower feed with slightly lower protein content, typically about 16–18%. This supports healthy growth without excessive weight gain. The feed should also include a boost in niacin level. Ducklings typically need two to three times more niacin than chicks. This helps to produce energy for rapid growth and to promote proper development.

Grower-Developer Feed (Weeks 8–18)

During the development stage, ducks need feed with moderate protein content about 14–16%. A reduction in the protein level at this stage is recommended, because too high of protein during growth and development can lead to wing deformity called angel wing. Proper nutrition at this stage is needed for strong bones and muscles.

Layer Feed (Weeks 19+ for Laying Ducks)

At 19 weeks, ducks that will be used for egg production should be switched to layer feed with about 16–18% protein. It contains added calcium for eggshell formation and other nutrients to produce strong, viable eggs.

Maintenance Feed (Weeks 19+ for Non-Laying Ducks)

Ducks not used for egg production can be fed a maintenance feed with lower protein content, about 14–16%, to maintain their health and weight without excessive weight gain.

Foraging

It is not as easy to raise ducks in small, confined areas, like you can with chickens, because ducks are so messy. To reduce the messiness, it is best to allow more space for ducks, if possible. Allowing your ducks to forage during a time of free-ranging is the best option for those who have enough acreage. The amount of time spent free-ranging depends on your overall objectives for the ducks.

Free-ranging ducks have the advantage of foraging and feeding on a variety of natural foods, which can make up a notable portion of their diet. The exact percentage of their diet that comes from foraging can vary depending on the available resources, the season, time allowed, etc.

If you are wanting to produce eggs, the amount of time spent free-ranging should be managed in order to control the percentage of their diet that comes from foraging. Foraging for ducks generally consists of the following.

Grasses and Weeds

Ducks will graze on various grasses, legumes, and weeds in their environment. This can include clover, dandelions, and other green plants.

Insects

Ducks are natural foragers and will actively search for various types of bugs that will provide them with essential protein.

Aquatic Plants

If ducks have access to ponds or water sources, they may feed on aquatic vegetation, such as waterweed, duckweed, and algae.

The percentage of their diet that comes from foraging will vary, but it's essential to ensure an overall balanced diet. Ducks should still have access to commercial duck feed to ensure they receive all the necessary nutrients. A general guideline is to offer commercial feed as the primary diet (about 70–80% of their daily intake) and allow them to forage for the remainder 20–30% to supplement their diet with natural foods.

It is important, however, to monitor the condition of the ducks and adjust their diet accordingly. If they're not getting enough nutrients from foraging, you may need to increase the percentage of commercial feed in their diet to meet their nutritional requirements. Symptoms of this could be a drop in egg production and/or weight loss. Additionally, it is really important to provide access to clean water at all times for proper health and digestion.

Flock Watering

Water consumption is important for your duck's performance and health. The things that affect water consumption the most are water quality and water availability. To optimize water quality and availability, you should carefully consider your water management and water equipment.

Water Management

Like chickens, water consumption is important for digestion, waste elimination, and body temperature. Optimum water intake will also help to optimize feed intake, which is necessary for maximum growth and egg production.

Duck water management involves ensuring they have access to clean and fresh water at all times. Ducks not only drink water but also use it for bathing and grooming.

Ducks need access to water deep enough to submerge their bills and heads to clear their nostrils and eyes. This water depth varies by duck species and age, but typically ranges from 4 to 8 inches.

Water temperature and water cleanliness is also important and is something that can be managed with proper water equipment and maintenance.

For a more thorough explanation of water management, refer to Flock Watering for Chickens in Chapter 1, Part 2.

Water Equipment

Cool, clean water should be freely accessible to your ducks at all times. Water helps to control body temperature and is important during hot weather.

Daily water intake volume will increase with age, so the size and number of waterers available will be an important consideration. On average, an adult duck will drink about 1 pint of water per day and up to 2 pints when it is hot.

If water consumption was the only criteria for choosing the size of a waterer, a 2-gallon waterer can service 6–8 ducks for a day. But because ducks will play, splash, and dip their heads in the water, they actually waste more water than they consume. The size, therefore, needs to be large enough to avoid having to refill it often.

Some homesteaders use large open metal or plastic bowls or pans for water. This provides more quantity and depth for head dipping, but the water will get dirty and will require frequent cleaning and replenishment with fresh water.

The most common types of waterers for ducks are the gravity flow/self-filling type and the automatic type. One way to avoid frequent filling is to have waterers that are on an autofill with float system.

To manage the watering area, it is often recommended to elevate the waterers and put a layer of gravel on the ground under the waterers to prevent the area from getting muddy and messy from all of the water splashing.

When making your waterer choices, you should consider which option best performs the following.

1. Provides cool water.
2. Promotes clean water.
3. Prevents algae formation.
4. Allows for heating to prevent freezing.

Water equipment choices can impact water management. Here are some common options.

Gravity Flow Waterers—Galvanized

The galvanized gravity flow waterers consist of a container and a corresponding round trough attached to the bottom of the container. The inner tank is filled with water from the top, and an outer shell fits over the inner tank and locked into place with a turning motion. Water is then allowed to fill the trough, and the trough water level is vacuum controlled.

Most galvanized gravity flow waterers are designed to sit on the ground, in lieu of being hung. Because the water can become dirty if placed directly on the ground, it is recommended to elevate the trough level to about 5–6 inches off the ground by placing the waterer on a small platform.

Being double walled, the galvanized waterer is able to keep the water relatively cool in the summer and free of algae. For wintertime freeze protection, the galvanized waterer can be placed on an electric heated base that is thermostatically controlled.

This type of waterer works well for chickens and can work for young ducks, as long at the trough at the bottom is large enough to allow the ducks to dip their entire head in the water.

Automatic Water Troughs

These are shallow troughs that are connected to a water source and provide a convenient way to give ducks continual access to fresh water. Most trough type delivery systems have a float valve for water level control.

They are efficient for larger flocks but require proper maintenance to ensure cleanliness. It is best to elevate the trough to better accommodate head dipping and prevent the issue of poop getting in the water.

The trough needs to be deep enough and open enough for the ducks to be able to dip their heads in the water.

Automatic Water Cups

The cup or bowl is connected to a water line and has a small flap, which when pushed by the duck's bill will allow water to flow into the cup. This option is great for providing clean water on demand.

Ducks learn to use them quickly. The best models are those designed for ducks and geese and, therefore, have large enough bowls to accommodate head dipping.

Drip System

For ducks in a free-range setting, a drip system can provide a continuous slow flow of water to allow for natural foraging.

Choose the water equipment based on the number of ducks, available space, and your specific management system. Be sure to regularly clean and maintain the water equipment in order to ensure the ducks have access to clean water at all times.

For a more thorough explanation of water equipment, refer to Flock Watering (for Chickens) in Chapter 1, Part 2.

Egg Production

Duck hens do not need a drake to lay eggs, but drakes are required for egg fertilization. Ducks will begin to lay eggs about 20–24 weeks of age.

On average, ducks can actually lay more eggs than chickens, and their eggs are about 30–50% larger. Large chicken eggs weigh roughly 24–26 oz. per dozen, while duck eggs, on average, can weigh 32–34 oz. per dozen. Duck eggs are higher in vitamins and omega-3 fatty acids than chicken eggs and stay fresher longer.

Ducks can consume 20–30% more feed per dozen eggs produced than chickens, but they can forage to meet some of their nutritional requirements. Ducks typically lay their eggs early in the morning, so they can be let out to forage mid-morning if necessary. Duck eggs can come in a variety of colors, ranging from light greens to light blues.

Egg Production Enhancement

Optimum egg production of your homestead will require the consideration and implementation of key factors, such as breed selection, housing and space, nesting, egg collection, nutrition, free-range options, lighting, cleanliness, health, and pest and predator control.

Best Breeds

Some breeds are best for egg production, and some breeds are used for both eggs and meat. Your breed choice will depend on your preferences and goals.

Housing and Space

A clean and well-ventilated shelter will be needed, with adequate space for each duck. Each bird should have at least 4 square feet of space in the shelter, and 14–16 square feet of space in the outdoor pen.

Nesting

Unlike chickens, ducks do not need nesting boxes to lay eggs. If boxes are available, they may or may not elect to use them. They are perfectly happy to make a nest in the bedding material and lay their eggs on the floor of the shelter. If you do choose to have nesting boxes in the shelter, they should be at floor level. The best nesting material in the boxes is wood chips or sawdust.

Egg Collection

Ducks usually lay their eggs early in the morning, so egg collection mid-to-late morning will usually suffice. Eggs should then be cleaned and promptly stored in the refrigerator.

Proper Nutrition

To maximize egg production, a balanced and high-quality duck layer feed that meets the nutritional needs of your ducks should be available as free access or fed by hand. A layer feed with at least 16–18% protein and 2.5–3.5% calcium is essential for good egg production. Fresh, clean water available at all times is also essential.

If the handfeeding option is followed, each duck should be fed 3.5 lb. per day divided into 60% in the morning and 40% in the afternoon.

Free-Range

If you have the space, it will be a great advantage to allow your ducks access to outdoor areas where they can forage for insects, grass, and other natural foods. This can not only help to supplement their diet but can also help to keep your barnyard less messy. For better egg accountability, some choose to allow free-ranging for a short time after eggs have been laid in the shelter in the morning.

Lighting

Ducks require a certain amount of daylight to trigger egg production. Ideally, the ducks need 15–17 hours of daylight each day in order to maintain a proper egg laying schedule. When natural daylight is insufficient, artificial lighting can be added to the shelter to achieve the needed daylight hours.

Cleanliness

To keep your ducks healthy and productive, you should maintain a regular routine of keeping the shelter and outside pen clean of poop and soiled organic material. One way to facilitate the outside duck pen cleanliness is to add a layer of sand. The sand will help in moisture absorption.

Keeping the water source clean will be an additional challenge. You should try to ensure that the water supply is clean and accessible.

Health

You should also maintain a vaccination and healthcare schedule for your ducks and consult a veterinarian when health issues arise.

Pest and Predator Control

You should protect your flock from pests and predators by elevating the shelter off the ground and securing the shelter and outdoor pen with strong fencing. Duck feed is a strong attractant to rodents. Larger rodents can eventually be attracted to the eggs as well. Infestations can be avoided with proper vigilance and the use of proper pest control measures.

Wintertime Egg Production

Ducks do not tend to have a wintertime reduction in egg production, as chickens do. However, there are a couple of precautions you can take to ensure adequate wintertime production.

Reduced Daylight

Ducks are influenced by the amount of daylight they receive. As the days get shorter in the fall and winter, artificial lighting will be required to maintain the needed 15–17 hours of light for optimum egg production.

Temperature and Weather

Cold temperatures can induce stress and cause the ducks to direct their energy toward staying warm instead of egg production. Extremely cold weather can also lead to frozen water sources, making it difficult for ducks to stay properly hydrated, which can also impact egg production.

Shelter

In mild climate areas, a three-sided shed can be used as a shelter, but in areas that have wet weather, cold winter weather, and/or lots of predators, an enclosed shelter should be used.

Your ducks should have a structure available that provides a safe area to sleep, lay eggs, and be protected from harsh weather and from predators. If enclosed, the shelter should include windows for natural lighting and ventilation, a floor covered with litter material for nesting and moisture absorption, and artificial lighting for controlling daylight hours.

Space Requirements

On average, ducks need about 4 square feet each inside the shelter. To use an example of a flock of 12 ducks, a structure of at least 48 square feet would suffice.

Nesting

Ducks do not need a roost for nighttime perching like chickens, nor do they require nesting boxes. If you have nesting boxes, they may choose to use them, but in order to do so, the boxes will need to be at floor level.

The ducks will be completely content to nest and lay eggs in the bedding material for nesting. With that in mind, you should consider the best type of bedding material.

Bedding Material

The floor will need to be covered with bedding material. The bedding material will serve two purposes: for nesting and for absorbing moisture. To meet the objective of these two purposes, you may need two different types of bedding material. For instance, you could have a bottom layer of sawdust or wood shavings for moisture absorption covered with an upper layer of straw for nesting. The bedding material will need to be replaced more often than you would for chickens.

Construction

The Chicken Coop section in Chapter 1, Part 2, described how to build a chicken coop for an example flock of 12 chickens. That same structure can be used for an example flock of 12 ducks, as the 48 square feet meets the minimum square feet per duck requirement.

The same dimensions and building material can be used. The only difference would be that neither a roost nor nesting boxes would be needed.

Like the chicken coop, the structure should be elevated to better protect the ducks from predators and from flooding when it rains, as the ducks will be nesting on the floor.

Ramp

Being elevated, a ramp will be needed, but the ramp for the ducks should have a more gradual incline than what is possible for chickens.

Electricity

The shelter needs to be equipped with electricity so that artificial lighting can be installed. Ducks need 15–17 hours of daylight each day to maintain a normal egg laying schedule. The lights can be set on a timer in order to achieve the desired hours of light.

The electricity can also be used to have heaters in the winter when temperatures fall into the 30s (F.), to prevent their eggs from freezing.

Feeding and Watering

Feed and water should be kept outside whenever possible. Because ducks are messier than chickens, having feed and water in the shelter will quickly lead to wet and soiled bedding.

Containment

In addition to the shelter, the ducks should also have an outside pen for containment. This is mainly for protection from predators and when they are not able to forage when the weather is bad.

Space Requirement

The required space per duck in an outside enclosure is 14–16 square feet per duck. For an example flock of 12 ducks, the outside square footage would need to be a minimum of 168–192 square feet. An outside pen that is roughly 24 ft. x 8 ft. producing 192 square feet would suffice.

The Chicken Run section in Chapter 1, Part 2, described an outside pen that was 16 ft. x 8 ft. The same method of construction and type of material can be used—it would just need to be 8 feet longer. The main reason for the extra space needed is because ducks are messier than chickens.

Unique Behavior

Ducks enjoy playing in water. It isn't mandatory to have a pool or a pond of water for them to swim and play in, but they certainly thrive when it is available.

The only problem is that they poop in the water. The water will become green and dirty quickly. Unless you are willing and able to keep a small pool clean, by dumping the water and cleaning the pool rather frequently, the only viable option is to have a relatively large pond for them to use. Refer to the Cost/Benefit section later in this chapter.

Healthcare

Common health issues in a small homestead flock of ducks can include the following.

Respiratory Infections

Ducks can suffer from respiratory infections due to poor ventilation or exposure to cold and damp conditions. Symptoms may include sneezing, nasal discharge, and labored breathing. The best way to avoid this is to ensure proper ventilation in the duck shelter, provide dry bedding, and protect the ducks from extreme weather conditions.

Parasites

Ducks can be affected by internal parasites, like worms, and external parasites, like mites or lice. Parasitic infestations can cause weight loss, lethargy, and feather loss. The best way to prevent this is by keeping the shelter clean, periodically deworming your ducks when necessary, and providing an adequate water source for bathing.

Botulism

Ducks are prone to *botulism*, which is a bacterial infection that can result from consuming contaminated water or food. Symptoms can include paralysis, difficulty breathing, and sudden death might result. The best way to avoid this problem is to provide a clean and reliable water source and avoid feeding ducks moldy or spoiled food.

Bumblefoot

This is a bacterial infection that affects a duck's feet, causing swelling and lameness. You can prevent this by providing clean and dry bedding, by keeping the duck area free from sharp objects that could cause injuries, and by promptly treating any wounds.

Egg-Related Issues

Ducks may experience egg-related problems, such as egg binding (difficulty laying eggs) or soft-shelled eggs. The best way to avoid this is to ensure your ducks have a balanced diet with adequate calcium, and by providing clean bedding material for nesting.

Duckling Issues

Newly hatched ducklings are vulnerable to a range of issues, including a common problem called pasty butt (blockage of the vent with feces) and leg deformities. This can be avoided by keeping the brooder clean and warm.

Dietary Problems

Ducks may develop nutritional deficiencies or overweight problems if not provided with a balanced diet. These types of issues can be avoided by feeding the ducks a balanced diet that is appropriate for their age and stage of growth and by having access to clean water at all times.

Cleanliness, appropriate and adequate feed, and clean water are mandatory for maintaining good health in a small homestead flock of ducks. If you suspect a health problem, you should consult a veterinarian who specializes in poultry for a proper diagnosis and treatment.

Expenses

The following expense data will be based on an example flock of 12 hens, beginning by purchasing a dozen ducklings from a hatchery.

Start-Up Expense Summary Charts

Start-up expenses will include the initial investment in ducklings and the cost for equipment and housing.

Initial Poultry Investment Expense Chart

Item	Cost
12 ducklings @ $11.00 each	$132

Initial Feed Investment to Get Ducklings to Laying Age

Starter Feed (20% Protein)

Ducklings will be fed duckling starter feed for 2 weeks. On average, each duckling will consume roughly .16 lb. per day for 14 days.

- .16 lb. per day for 14 days = 2.2 lb. total per duckling for 2 weeks
- Average price of $29 for 25 lb. = $1.16 per lb.
- 2.2 lb. @ $1.16 per lb. = $2.55 per duckling for first 2 weeks

Grower Feed (18% Protein)

Ducklings should be fed grower feed for 6 weeks (from week 3 to week 8). For this 6-week period, the ducklings will consume an average of .25 lb. per day of grower feed.

- .25 lb. per day for 42 days = 10.5 lb. total per duckling for 6 weeks
- Average price of $33 for 50 lb. = $0.66 per lb.
- 10.5 lb. @ $0.66 per lb. = $6.93 per duckling for 6 weeks

Grower-Developer Feed (16% Protein)

Ducklings should be fed grower-developer feed for 10 weeks (from week 9 to week 18). For this 10-week period, the ducklings will consume an average of .35 lb. per day of grower feed.

- .35 lb. per day for 70 days = 24.5 lb. total per duckling for 10 weeks
- Average price of $36.60 for 50 lb. = $0.73 per lb.
- 24.5 lb. @ $0.73 per lb. = $17.88 per duckling for 10 weeks

Item	Cost per Bird	Cost for 12
Feed cost for weeks 1–2	$2.55 per duckling	$30.60
Feed cost for weeks 3–8	$6.93 per duckling	$83.16
Feed cost for weeks 9–18	$17.88 per duck	$214.56
Total	$27.36	$328.32

Infrastructure and Equipment Expense Chart

Item	Cost
Brooder box	$45
Heat lamp for light	$18
Brooder heat plate	$45
Duck feeder & waterer set	$18
Hen feeders	$25
Hen waterers	$35
Duck shelter—($13.50/sq. ft.)	$650
Duck pen—($2.00/sq. ft.)	$384
Total	**$ 1220**

Operating Expenses

Operating expenses will include layer feed for a small flock of 12 laying hens. This example will not include a drake.

Feed Costs

Laying hens will eat, on average, about .35 lb. of feed per day. Layer feed can be purchased for $21 for a 25-lb. bag. The following chart will be feed cost for 12 hens.

- .35 lb. per hen for 12 hens = 4.2 lb. feed per day
- 4.2 lb. for 7 days = 29.4 lb. for 12 hens per week
- Average price of $21 for 25 lb. = $0.84 per lb.
- 29.4 lb. @ $0.84 per lb. = $24.70 for 12 hens per week

Item	Daily Cost	Weekly Cost
Layer feed for 12 hens	$3.53	$24.70

Animal Care

Bedding

Wood shavings can come in bales of 8 cu. ft. Each bale costs roughly $8. One bale is needed to cover the floor of a 48 sq. ft. coop, 2 inches deep. The bedding should be replaced every 2–4 weeks. If the bedding is replaced every 2 weeks, the weekly cost would be $4. One straw bale will cost roughly $5 and will last for about 2 weeks as a nesting layer on the floor.

Vet Care

Vet care can vary, but the prevailing cost estimate is $52 per year for a dozen hens.

Item	Weekly Cost
Shelter bedding (straw bale)	$2.50
Shelter bedding (wood shavings)	$4.00
Vet care	$1.00
Total	**$7.50**

Cost/Benefit

The purpose for this analysis is to provide a way to examine the cost versus the benefit of raising ducks for household eggs. The analysis will compare the weekly costs for egg production to the cost of purchasing eggs from a store. Please note that operating costs only are used in this analysis—start-up costs are not included.

Savings/Loss

On a weekly average, ducks can lay more eggs than chickens. Using a figure of 6 eggs per hen per week, you could theoretically get 6 dozen eggs per week, with an example flock of 12 hens.

Because duck eggs are 30–50% larger, the value of 6 dozen eggs per week, based on local prices, would be roughly $5.50 per dozen.

- **Note:** The price per dozen is an estimate and can vary depending on the locale and the economy.

Item	Weekly Expense	Weekly Value or Revenue	Weekly Savings/ Loss
Feed & misc.	$24.70		
Animal care	$7.50		
Total cost for 6 doz. eggs	**$32.20**		
Egg value per 6 doz. eggs		$33.00	
Total value/revenue		**$33.00**	
Savings/loss			**$.80**

As you can see, the cost/benefit for raising your own eggs is a near break-even proposition, and this isn't taking into consideration the start-up costs and value of labor. Based on the data above, the cost per dozen is about $5.37, which is about $0.13 less per dozen than purchasing duck eggs at the store.

The major benefit in producing your own eggs is twofold. The quality of the eggs is probably better, and you also have the enjoyment and the peace of mind of being self-sustaining and raising your own food off the grid.

Cost/Benefit Adjustments

The number of eggs you get from ducks vs. chickens is 30–50% more per week, the size of duck eggs is 30–50% larger, and thus the price per dozen is about 30–50% higher. However, ducks eat 30–50% more than chickens, and the cost of feed is about 30–50% more. The end result is thus: like raising chickens for eggs, the cost/benefit for raising ducks for eggs (using commercial feed) is about even.

With that said, if you are looking for the best option for eggs, and you have limited space, raising chickens would be a wiser option, because you can raise chickens in a smaller confined space without the messiness you get with ducks.

However, if you have enough available space for the ducks to forage on a grassy area adjacent to the shelter and the pen, this would make the option of raising ducks more feasible. In addition, if you also have a reasonable-sized pond, then you truly have a winnable scenario for raising ducks.

Please note that foraging can reduce the need for commercial feed by 20–30% of the total daily diet requirement. This supplement option can reduce the overall cost of production, thus making the savings/loss figure more profitable and the cost/benefit more feasible.

In regard to adequate space for a pond, the optimum size of a pond for 12 ducks should be about 225 square feet (18.75 sq. ft. per duck) with a minimum depth of 3–4 ft. deep. To help in keeping the pond reasonably clean, you should have an aerator in the center of the pond.

You might even want to consider having carp in the pond to help with algae control. The minimum depth of 3–4 ft. will help to maintain appropriate water temperatures and oxygen levels in order to make a fish option possible.

It is also possible to benefit from a pond fertilized with duck poop by arranging for the pond to be at a higher elevation than a garden. The water could theoretically be piped to the garden, by gravity feed, for water high in nitrogen. This, of course, would create a need to fill the pond at the same rate it is being piped out to the garden.

If the property is somewhat level, the pond water could be pumped to the garden. To save on energy costs, power for the pump could be provided by a small solar panel. Again, pond water would need to be replenished at the same rate as water pumped out to the garden.

Picture Gallery

Egg Ducks & Meat Ducks

Khaki Campbell

Indian Runner

Welsh Harlequin

Ancona

Pekin

Roen

Dual Purpose Ducks

Cayuga

Silver Appleyard

Saxony

Buff Orpington

Magpie

Blue Swedish

CHAPTER 3
Geese

If you have the acreage for foraging/grazing geese and can put up with their unique behavior, the decision to raise geese for meat can be a no-brainer. With a reasonable-sized pasture of grass for the geese to graze on, you can easily produce a notable amount of meat without much expense.

Geese are also known for their weeding capability. You can use geese to weed your own fields, or they can be possibly hired out to weed neighboring fields. They can eat grass, weeds, insects, and even harmful invertebrates, such as slugs and snails.

In this chapter, the most common breeds for meat and for weeding are discussed in addition to feeding, grazing, watering, meat production, and meat processing. Weeding management will be evaluated and information provided regarding shelter and containment requirements. The more common health issues will be listed and how to avoid them.

The chapter concludes with a cost/benefit analysis to compare the cost of household meat production to the value of comparable goose meat prices.

Breeds

Any goose breed can be used for meat and weeding, but some breeds are more commonly used for these purposes. There are three breeds of geese that are primarily used for meat because of their fat content and quality of meat. There are two breeds commonly used for both meat and weeding,

Breeds for Meat

The three breeds preferred for their quality and quantity of meat are listed below.

Embden

The Embden geese originated from Germany. Their plumage is pure white with an orange bill and orange feet. The adult Embden geese can weigh between 18 and 25 pounds. The Embden geese are generally friendly and easy to handle. They are known as a popular choice for meat production on small farms.

Embden geese are known for producing high-quality meat. The meat is juicy, tender, and has a mild flavor, making it suitable for a variety of culinary uses.

Toulouse

The Toulouse geese originated from France. Their plumage is grayish blue with a white underbelly, orange bill, and orange feet. An adult Toulouse goose can weigh between 18 and 22 pounds. The Toulouse geese are known for their calm and docile temperament, making them suitable for small farms.

Toulouse geese are renowned for their excellent meat quality. The meat is tender and flavorful and is typically used in traditional French dishes.

American Buff

The American Buff originated from the United States. Their plumage is a light buff or fawn color with orange bills and feet. The typical adult weight is between 12 and 16 pounds. American Buff geese are known for their calm and gentle disposition. They are easy to manage and are suitable for small-scale meat production.

American Buff geese are valued for their high quality, well-flavored meat. It is considered to have a good balance of tenderness and juiciness.

Summary

All three of these goose breeds are recognized for their meat quality with variations in flavor and texture to suit different culinary preferences.

Dual-Purpose Breeds

There are two breeds highly regarded for their weeding abilities that can also serve a dual purpose for meat.

Chinese

The Chinese geese originated in China. Their plumage is generally white with orange bills and legs. The typical adult weight is between 10 and 14 pounds. These are known for their hardiness and adaptability. They have lean meat and typically make good foragers, suitable for weeding on small farms.

African

The exact origin of the African goose is uncertain, but they have been bred in the United States and other regions. Their plumage is typically gray with a white head, orange bill, and orange legs. Adult African geese generally weigh between 10 and 14 pounds. The African geese are known for their alert and protective nature. They can be a bit more aggressive than some other breeds, especially when protecting their territory. They are efficient foragers and weeders.

Other Breeds

There are three other breeds that can also be used as dual-purpose birds, but to a lesser degree: Roman Tufted, Sebastopol, and Pilgrim.

Breed Meat Yield Comparison

Criteria	Embden	Toulouse	American Buff	Chinese	African
Live wt. (lb.) @ 16, 18, & 20 weeks	16, 18, & 20 lb.	16, 18, & 20 lb.	12, 14, & 16 lb.	10, 12, & 14 lb.	10, 12, & 14 lb.
Processed wt. (lb.) @ 16, 18, & 20 weeks	12, 14, & 16 lb.	12, 14, & 16 lb.	8, 10, & 12 lb.	7, 9, & 11 lb.	7, 9, & 11 lb.

Hatching and Brooding

The incubation and brooding process is similar to that of ducks. For a detailed description of the process and various equipment options, refer to Hatching and Brooding (for Ducks) in Chapter 2, Part 2.

Incubating Eggs

Basically, the same incubator and method of use can be used for both ducks and geese. The only difference for geese eggs is the number of days required for hatching. Goose eggs require, on average, 28–35 days to hatch. Geese eggs should be incubated at a temperature of 99.5 degrees F. until pipping.

Humidity should be 55% for the first 27 days and the eggs turned four times per day. On day 28, egg turning should stop, and the humidity should be increased to 75% until pipping starts (day 28–35). When pipping begins, the humidity should be increased to 85%, and the temperature be reduced to 98–99 degrees F.

Brooders

The process and equipment for brooding goslings are similar to that for ducklings. The main differences are temperature and space required. Goslings will remain in the brooder for 6–8 weeks.

Brooder Space

The space requirement per gosling is about 2 sq. ft. per for the first 3 weeks. For weeks 4–8, each gosling will require about 4–6 square feet to reduce the rate at which bedding is soiled.

Brooder Heat

The temperature should begin at 90 degrees F. and be reduced 5–10 degrees per subsequent week thereafter until 70 degrees F. is reached. Like for ducklings, the heat plate works best as a heat source, as it allows for daylight hours management. The light management would be the same as for that for ducklings.

Brooder Water

Because goslings tend to be messy and splash the water, a special accommodation is required. Goslings like to splash and dip their heads in the water. This creates quite a wet mess around the waterer area. To accommodate this, a good brooder choice can be the galvanized metal livestock water trough that is rectangle in shape. The size can be 2 ft. tall x 2 ft. wide x 4 or 6 ft. long.

For water management, you can also place a metal cookie baking pan with a ¾-inch lip on the opposite end of the brooder from the feeder. You can fill the pan with gravel, and then place the waterer on top of the gravel. This will help to contain the splashed water and assist in preventing the bedding from getting too wet. You will need to clean the pan and change the water frequently. The same type of waterer that is used for ducklings will work fine.

Brooder Bedding

Gosling poop contains a lot of moisture. For that reason, bedding can become moist and messy quite frequently. With that in mind, wood pellets can be a better alternative for bedding than wood shavings. The wood pellets will absorb the moisture much better and last longer than wood shavings.

The top of the brooder should be covered to protect the goslings. Typically, the top would consist of a metal frame covered with hardware cloth or chicken wire. The most common predators for brooders are cats.

Brooder Feeding

During the first 2 weeks, each gosling will consume about .2 pounds of feed per day. Between weeks 1–7, the feed consumption per gosling will gradually increase from .2 lb. to .35 lb. of feed per day. During their stay in the brooder, the goslings should have continuous access to feed and water.

The same bulk type of feeder that is used for ducklings will work for goslings in the beginning. After about 3–4 weeks, you may need to switch to a metal trough type feeder, so that the goslings can have full access to feed at all times.

As a feed choice, baby goslings should be fed starter crumbles, which is specially formulated for their growth and development. Goslings need more protein than adult geese, especially in the first 2 weeks of their lives. The gosling starter feed has a protein content of 20–22%, which is ideal for the first 2 weeks. For weeks 3–7, you can reduce protein content to roughly 16–18%. Gosling grower feed can be used for weeks 3–7 in a small pellet form with higher niacin content.

Flock Feeding

Geese will be raised primarily for meat and weeding, so they will be introduced to foraging much earlier than ducks. Foraging will become a major part of their daily diet during summer months and be supplemented with commercial feed during winter months.

Commercial Feeds

The most common feeds for geese include commercial starter, grower, and maintenance feeds. Feeding geese requires specific types of commercial feed that are formulated to supply their nutritional needs as they grow and develop.

Starter Feed (Weeks 1–2)

Newly hatched goslings should be fed a high-protein starter feed containing about 20–22% protein. This helps promote early growth and feather development.

Note: It is not recommended to feed your goslings medicated chick starter. Goslings eat more than chicks and can overmedicate themselves. Chick feed is medicated to prevent coccidiosis, which does not typically affect goslings.

Grower Feed (Weeks 3–18)

As the goslings grow, you will need to transition them to a grower feed with slightly lower protein content, typically about 16–18%. This supports healthy growth without excessive weight gain. The feed should also include a boost in niacin level. Goslings and ducklings typically need two to three times more niacin than chicks. This helps to produce energy for rapid growth and to promote proper development.

Maintenance Feed (Weeks 19+ for Non-Laying Geese)

Geese raised to 20–24 weeks for meat production can be fed a maintenance feed with lower protein content, about 14–16%, to maintain their health and weight without excessive weight gain.

Grazing

Grazing is going to be a major part of the goose diet. You won't want to try to raise geese without this availability.

Allowing geese to forage and graze provides the advantages of feeding on a variety of natural foods, like weeds, grass, and insects, which will make up a notable portion of their diet. Geese can begin foraging at an earlier age than other types of poultry.

Unlike ducks that use commercial feed for 70–80% of their diet and foraging for the balance, geese are just the opposite. In fact, it is not uncommon for geese to use foraging for 100% of their diet during the summer months and use commercial feed to make up most or all of their wintertime diet when grass is scarce.

Unlike ducks and chickens that will prefer commercial feed to foraging, geese will often not pursue commercial feed at all, if plenty of grass is available. This makes raising geese practical for meat and weeding, if you have the acreage.

Grasses and Weeds

Geese will graze on various grasses, legumes, and weeds in their environment. This can include pasture grass, lawn grass, clover, dandelions, and a host of other green plants.

Insects

Geese are natural foragers and will actively search for various types of bugs that will provide them with essential protein.

Slugs and Snails

While foraging, geese are known to eat harmful invertebrates, such as slugs and snails that are often a problem in orchards and vineyards.

Flock Watering

Water consumption is important for geese performance and health. The things that affect water consumption the most are water quality and water availability. To optimize water quality and availability, you should carefully consider your water management and water equipment.

Water Management

Like ducks, water consumption is important for digestion, waste elimination, and body temperature. Optimum water intake will also help to optimize feed intake, which is necessary for maximum growth and meat production.

Geese water management involves ensuring they have access to clean and fresh water at all times. Geese not only drink water but also use it for bathing and grooming.

Geese need access to water deep enough to submerge their bills and heads to clear their nostrils and eyes. This water depth varies by geese species and age but typically ranges from 4 to 12 inches.

Water temperature and water cleanliness is also important and is something that can be managed with proper water equipment and maintenance.

For a more thorough explanation of water management, refer to Flock Watering (for Chickens) in Chapter 1, Part 2.

Water Equipment

Cool, clean water should be freely accessible to your geese at all times. Water helps to control body temperature and is important during hot weather.

Daily water intake volume will increase with age, so the size and number of waterers available will be an important consideration. On average, a goose will drink about 2–4 pints of water per day, depending on the outside temperature.

Because geese will play, splash, and dip their heads in the water, they will tend to waste more water than they consume. The size of waterer, therefore, needs to be large enough to avoid having to refill it often.

Some homesteaders use large metal or plastic tubs for water, ~8 inches deep. This provides more quantity and depth for head dipping, but the water will get dirty and thus will require frequent cleaning and replenishment with fresh water.

The most common types of waterers for geese are the open tubs or troughs and the automatic waterer type. One way to avoid frequent filling is to have waterers that are on an autofill with float system.

To manage the watering area, it is often recommended to elevate the waterers and put a layer of gravel on the ground under the waterers to prevent the area from getting muddy and messy from all of the water splashing.

When making your waterer choices, you should consider which option best performs the following.

1. Provides cool water.
2. Promotes clean water.
3. Prevents algae formation.
4. Allows for heating to prevent freezing.

The choice of water equipment can impact water management. Here are some common options.

Open Tubs or Troughs

Large, open tubs placed at ground level can work well for water access. Ensure they are kept full and clean, so the geese can dip their heads in to drink.

Small troughs, 6–8 inches deep, designed for livestock feeding troughs can also be used for geese. They are durable and designed to hold a significant amount of water. Make sure to keep them clean and free of debris.

Automatic Water Troughs

These are shallow troughs that are connected to a water source and provide a convenient way to give geese continual access to fresh water. Most trough type delivery systems have a float valve for water level control.

Because these automatic trough waterers are only 4–6 inches deep, it is best to elevate the trough to better accommodate head dipping and prevent the issue of poop getting in the water. The trough needs to be deep enough and open enough for the geese to be able to dip their heads in the water.

Automatic Water Cups

The cup or bowl is connected to a water line and has a small flap, which when pushed by the goose's bill will allow water to flow into the cup. This option is great for providing clean water on demand.

Geese learn to use them quickly. The best models are those designed for ducks and geese, and, therefore, have large enough bowls to accommodate head dipping.

Drip System

To accommodate foraging, a drip system can provide a continuous slow flow of water. You should choose the water equipment based on the number of geese, available space, and your specific management system. Be sure to regularly clean and maintain the water equipment in order to ensure the geese have access to clean water at all times.

Weeding Management

Generally speaking, geese will consume weeds as part of the overall foraging program, but it is also possible for geese to be used specifically and exclusively for weeding a particular area. For instance, my grandfather used a flock of geese to weed his cotton fields on our 400-acre farm.

Given that labor and fuel are becoming more costly for field weeding and cultivation, the prospect of using geese for weeding may be a viable option. It is possible to hire out your flock for weeding a field much like hiring out a herd of goats for brush control. The best breeds for weeding are the Chinese and African geese.

For foraging, you want to know how many geese your particular pasture can support for a given amount of time. But weeding is different—you are generally wanting to eliminate the weed problem as soon as possible, to prevent the weeds from growing and going to seed.

When and Where

The best time to weed is in the spring, before weeds and grass get too high. The common customers for geese weeding are farmers with row crops, plant nurseries, orchards, and vineyards. Geese can also be used to maintain fence rows and ditch banks, where hoeing or cultivating is difficult.

How Many

You probably would want to charge by the acre, but it is difficult to ascertain how many geese you need for a weeding project, and how long it would take because the extent of the weed problem can vary. As a general rule of thumb, you can use 4–6 geese per acre for row crops with an early low-to-moderate grass and weed problem.

Weeds and Grass

Geese will eat a variety of weeds, such as dandelion, bindweed, curly dock, and lamb's quarters. They also readily consume troublesome grasses, such as Johnson grass, Bermuda grass, sedges, watergrass, nut grass, crabgrass, clover, and chickweed. Your selling point is twofold: the flock is not only eliminating unwanted grass and weeds, but also is helping to fertilize the ground at the same time.

Timing

The denser the grass and weed problem, the more geese you will need, or the longer time it will take to finish. Be diligent to move your geese to the next area, when one area is clean of grass and weeds. If you leave them too long, they may begin to feed on the desired plants the customer is trying to protect.

Fencing and Shelter

You may need to supply your own fencing if the target area isn't fenced. It isn't difficult to keep the geese in a particular area if there is plenty to graze on. If you do need to set up temporary barriers, you can use a mesh fencing made of nylon or plastic for a reasonable price. You can get these in 4-ft. x 100-ft. rolls for under $40.

If the target field is not on your property, you will also need to arrange for a portable enclosed pen or shelter to contain and protect the flock each night. This is important for protection from predators until the job is done.

Water and Supplemental Feed

Water will need to be supplied during the day at both ends of the field. Periodic water movement can facilitate geese movement for wider weeding coverage. It is a good idea to have goose maintenance feed available for the geese at the end of the day as well. The feed can help to gather your flock for penning in the evening.

Meat Production

Raising geese for meat is a common strategy, but before proceeding with a plan, a few things should be considered, such as numbers, processing ages, and plans for the meat (selling vs. home use).

Best Processing Ages

Part of the processing procedure includes the decision to either pluck the feathers or skin the carcass. Most choose to pluck the feathers and leave the skin on. The skin adds fat and flavor and helps to keep the meat juicy when cooking. Skinning, on the other hand, is much faster. Most homesteaders who are contemplating selling some or all of the processed meat elect to pluck the fowl to facilitate sale preferences.

Geese and ducks have a lot of feathers, and these feathers are replaced periodically with new feathers. These new feathers begin as pin feathers, which are extremely hard to pluck. With that in mind, it is best to choose certain weeks of growth, when the pin feathers can be avoided. For geese, these optimum ages are 9, 15, and 20 weeks.

These ages coincide with the optimum ages for meat production, which are 16, 20, and 24 weeks. It is more optimal to graze a few more geese, then process earlier at 16 weeks.

Production Management

Before you begin raising geese for meat, you should verify that you have two things: (1) an adequate amount of acreage with grass for foraging and (2) a plan for numbers and ages for processing.

The following explanation is an example of how you might assess your grazing potential, formulate a plan, and implement the plan.

Numbers and Weeks

Assume that you have 1 acre of grass. To find how many geese you can forage on the 1 acre, use the method described in Pasture Management (for Milk Cows) in Chapter 1, Part 1, which explains how to determine carrying capacity. Even though that method is generally used for livestock, you can still use that method to give an idea of how many geese your 1 acre will support.

After using the method to determine the carrying capacity, you can use an example result of 1 AUM on the 1 acre of grass for example data. You will now need to find the AUM equivalent (AUE) for geese, by dividing the 1 AUM by .012.

By using this formula, you will determine the total number of geese your 1 acre will support in the following example. You will also consider the best ages for processing and compare two batches: one for 16 weeks and one for 20 weeks to see how the meat production compares.

- 1 acre carrying capacity of 1 AUM—AUE for geese is .012.
- 1 AUM ÷ .012 = 83 geese for 1 acre for 1 month
- 1 summer batch for meat for 4 months (16 weeks)
  - 83 ÷ 4 summer months = 20 geese
- 1 summer batch for meat for 5 months (20 weeks)
  - 83 ÷ 5 summer months = 16 geese

By using the chart in the Breeds section in this chapter, you will use the processed weights for Embden geese at 16 weeks and 20 weeks.

- Processed weight at 16 weeks = 12 lb. x 20 geese = 240 lb.
  - 240 lb. x $9.50 per lb. = $2280 value
- Processed weight at 20 weeks = 14 lb. x 16 geese = 224 lb.
  - 224 lb. x $9.50 per lb. = $2128 value

By the way, if you were to raise a batch at 24 weeks, the result would be the same as 20 weeks—just fewer geese at heavier processing weight but the same total (224 lb. meat).

- Processed weight at 24 weeks = 16 lb. x 14 geese = 224 lb.

Meat Marketing

Before you decide to raise geese for meat to sell, you will need to verify that you have a market for your goose meat. There are several places where you might be able to market your meat. If you don't have a viable option for marketing your meat, you can always optimize your home use.

Farmers' Markets

Many homesteaders sell their products, including goose meat, at local farmers' markets. These markets provide an opportunity to connect directly with consumers who are looking for locally raised, high-quality products.

Local Butcher Shops

Contacting local butcher shops or meat processors may be a good option if you don't want to handle the processing and packaging of the goose meat yourself.

Food Coops

Local food cooperatives often work with small-scale producers to offer a variety of farm-fresh products to their members.

Local Homestead Network

Some homesteaders are able to sell directly to neighboring customers through a local farm customer network.

Home Use

You may not be able to find a viable market locally to sell your goose meat. If that is the case, you can still raise geese for your own meat, but you may want to adjust the number of geese you raise. Another possibility would be to raise the optimum number as per what your pasture will support and just freeze the extra meat for wintertime home use.

Meat Processing

If you choose to butcher your own geese, there are a few options to consider regarding dispatching, plucking, butchering, storage, etc.

Dispatching

Dispatching begins by suspending the goose's head down into a metal or plastic funnel that is attached to an upright support. The funnel prevents the bird from moving and allows the head to protrude from the bottom. While holding the goose's head in one hand, the jugular vein is severed on the underside of the neck, just behind the head.

Plucking

There are four common ways to pluck your goose: (1) dry method, (2) freezing method, (3) scalding method, and (4) auto machine method. The two most commonly used methods, which are the scalding method and the machine method, will be discussed.

Scalding Method

This method is fairly simple. You will need a pot large enough to totally immerse the fully feathered dead goose in hot water. A pot this size will need to be about a 5-gallon size. Fill the pot with water and bring it to a boil—about 212 degrees F.

By holding the goose by the legs, dunk the goose in the water (totally immersed), several times for about 1–2 minutes. Don't hold the carcass in the water longer than 2 minutes, as this will begin to cook the meat. After dunking for a couple of minutes, try plucking a few feathers to verify that the scalding has done its job.

After the scalding is complete, you may want to don rubber gloves so that you can begin the plucking right away without getting burned. Lay the goose on a cutting board, and begin to pull the feathers out, a few at a time. The scalding process should allow you to pluck the feathers easily.

Place the plucked feathers in a trash bag. Once all of the feathers have been removed, you are ready to start the butchering process.

Machine Method

The automatic poultry feather-plucking machine works amazingly well. It looks like an old-fashioned washtub, with rubber fingers on the inside wall and bottom. All you have to do is place the scalded bird in the tub and turn it on. You may want to spray water into the tub as it is turning to facilitate the feathers dropping out the bottom.

After the scalded goose has been placed inside the tub and the process started, it will be totally plucked clean in a minute or less. The feathers fall out the bottom onto the ground. It is truly amazing and saves so much time and effort. If you plan on raising geese for meat, and will have several birds to pluck at a time on a regular basis, this is the way to go.

Butchering

You can begin by severing the neck and head, the wings, and the legs at the knees. Then lay the goose carcass on its back on the cutting board. At the rearward end, make an opening between the tail and the breasts and remove the internal organs and entrails.

Wash the carcass thoroughly and prepare the carcass for storage in the refrigerator. Once it has been cooled for 3–4 hours, you can either package the goose for sale, prepare it for cooking, or wrap it for freezing.

Storage

If the scalding method is used for plucking, the shelf life will be diminished, and the goose will need to either be sold, cooked, or frozen soon.

Goose meat can be kept in the refrigerator for 1–3 days for home use. For sale, the goose can be vacuum sealed and can last 3–5 days in the refrigerator. When properly wrapped for freezing, the meat can last 6–12 months.

Shelter

In mild climate areas, a three-sided shed can be used as a shelter, but in areas that have wet weather, cold winter weather, and/or lots of predators, an enclosed shelter should be used.

Your geese should have a structure available that provides a safe area to sleep and be protected from harsh weather and from predators. If enclosed, the shelter should include windows for natural lighting and ventilation and floor covered with litter material for moisture absorption.

Space Requirements

On average, geese need about 6–8 square feet each inside the shelter. The indoor space should be connected to adequate outdoor space for grazing and/or exercise.

Nesting

Even though geese will lay eggs, their egg laying is seasonal, and they only lay 20–40 eggs per year. For that reason, most homesteaders don't raise geese for eggs. In the spring, geese hens will make a nest out of the straw on the floor of the shelter and will lay their eggs there. No special accommodation need be taken for egg laying other than the required sawdust and straw on the floor of the shelter for moisture absorption.

Bedding Material

The floor will need to be covered with bedding material. The bedding material will serve mostly for absorbing moisture. Because geese poop has such a high moisture content, you may need two different types of bedding material. For instance, you could have a bottom layer of sawdust or wood shavings for moisture absorption covered with an upper layer of straw for extra dry bedding material. The bedding material will need to be replaced more often than you would for chickens.

Construction

In the Chicken Coops section in Chapter 1, Part 2, instructions were given on how to build a chicken coop for an example flock of 12 chickens. That same type of structure can be used for an example flock of 12 geese, but the structure size would need to be double—a total of 96 square feet in order to meet the minimum square feet per goose requirement.

For this example, the dimensions and the building material can be doubled. The only difference would be that neither a roost nor nesting boxes would be needed.

Like the chicken coop, the structure should be elevated to better protect the geese from predators and from flooding when it rains, as the geese will be sleeping on the floor.

Ramp

Being elevated, a ramp will be needed, but the ramp for the geese should have a more gradual incline than what the chickens need.

Electricity

Geese are raised primarily for meat and/or weeding, so it is not mandatory to have electricity installed in the shelter for electric lighting. The lighting is a requirement for poultry that are being raised for eggs, which need 15–17 lighted hours each day.

Feeding and Watering

Feed and water should be kept outside whenever possible. Because geese are messier than chickens, having feed and water in the shelter will quickly lead to wet and soiled bedding.

Containment

In addition to the shelter, the geese should also have an outside pen for containment. This is mainly for protection from predators, and when they are not able to forage when the weather is bad.

Space Requirement

The required space per goose in an outside enclosure is 20 square feet per goose. For an example flock of 12 geese, the outside square footage would need to be a minimum of 240 square feet. An outside pen that is roughly 30 ft. x 8 ft. producing 240 square feet would suffice.

In the Chicken Run section in Chapter 1, Part 2, an outside pen that was 16 ft. x 8 ft. was described. The same method of construction and type of material can be used for geese—it would just need to be twice as long. The main reason for the extra space is because geese are messier than chickens and larger than ducks.

Unique Behavior

Geese don't need to have a water pond, like ducks enjoy. They are perfectly content with grazing and pooping. They do have a behavior that is somewhat unique to geese, however, and that is aggressiveness. Geese have a tendency to be aggressive and will bite, if they feel you are getting too close to their goslings.

Geese can also be quite loud—much more so than ducks. If you can adapt to these two things, and you have enough grass for them to graze, geese can be a viable option for inexpensive meat.

Healthcare

Raising geese can be a rewarding experience, but they can be susceptible to various health issues, like any other poultry on a homestead. Here are some typical health issues that a small flock of geese on a homestead might encounter.

Respiratory Infections

Symptoms can include sneezing, coughing, and nasal discharge. The typical causes are poor ventilation and being crowded in the shelter. This can be prevented by having adequate space, by replacing the bedding frequently, and by providing good ventilation.

Botulism

Symptoms for *botulism* can be signs of weakness and inability to walk or stand. The main causes are from ingesting contaminated feed or water. The best way to prevent this is to ensure their food and water is fresh and the dispensers are clean.

Parasites

The main symptoms from parasites are weight loss and diarrhea. The causes are basically internal or external parasites. The best prevention is regular deworming and maintaining a clean shelter.

Nutritional Deficiencies

The typical symptoms for this type of health issue are poor growth and feather abnormalities. The obvious cause is inadequate nutrition, which can be prevented by providing a balanced and appropriate diet.

Avian Influenza

The symptoms for avian flu are basically respiratory distress. The main cause is from contact with infected birds. The best prevention is implementing and maintaining a healthy protocol that includes monitoring for signs of illness on a regular basis.

Heat Stress

The symptoms of heat stress include panting and lethargy. The typical causes are high temperatures and inadequate shade. The best prevention is to provide shade, ensure the geese have access to cool water all day, and avoid overcrowding.

Gastrointestinal Issues

The main symptoms include diarrhea and decreased appetite. The main cause is contaminated food or water. The best prevention is to provide clean food and water sources.

Most health issues can be avoided by maintaining regular health checks, a clean living environment, and a well-balanced diet. If you notice any signs of illness, it's important to consult with a veterinarian who has experience with poultry.

Expenses

The following expense data will be based on an example flock of 12 Embden geese, beginning by purchasing a dozen goslings from a hatchery.

Start-Up Expense Summary Charts

Start-up expenses will include the initial investment in goslings and the cost for equipment and housing.

Initial Poultry Investment Expense Chart

Item	Cost
12 goslings @ $19.95 each	$239.40

Initial Feed Investment for Goslings

The feed investment for this example will be for the first 7 weeks while in the brooder. Ideally, the goslings should be transferred to a grass pasture where they will be allowed to forage daily, until they are old enough to butcher.

Starter Feed (20–22% Protein)

Goslings will be fed gosling starter feed for 2 weeks. On average, each gosling will consume roughly .20 lb. per day for 14 days.

- .20 lb. per day for 14 days = 2.8 lb. total per gosling for 2 weeks
- Average price of $29 for 25 lb. = $1.16 per lb.
- 2.8 lb. @ $1.16 per lb. = $3.25 per gosling for first 2 weeks

Grower Feed (16–18% Protein)

Goslings should be fed grower feed for 5 weeks (from week 3 to week 7). For this 5-week period, the goslings will consume an average of .35 lb. per day of grower feed.

- .35 lb. per day for 35 days = 12.25 lb. total per gosling for 5 weeks
- Average price of $54 for 50 lb. = $1.08 per lb.
- 12.25 lb. @ $1.08 per lb. = $13.23 per gosling for 5 weeks

Item	Cost per Bird	Cost for 12
Feed cost for weeks 1–2	$3.25 per gosling	$39.00
Feed cost for weeks 3–7	$13.23 per gosling	$158.76
Total	**$16.48**	**$197.76**

Infrastructure and Equipment Expense Chart

Item	Cost
Brooder box	$45
Heat lamp for light	$18
Brooder heat plate	$45
Gosling feeder & waterer set	$18
Gosling trough feeder	$23
Hen feeders	$25
Hen waterers	$35
Goose shelter — ($13.50/sq. ft.)	$1296
Goose pen — ($2.00/sq. ft.)	$480
Total	**$ 1985**

Operating Expenses

Unlike for chickens and ducks, the operating expenses will not include layer feed for laying hens, because the geese are being raised for meat and will be allowed to forage during the day, in lieu of being kept in an outside pen.

The operating expenses will, therefore, be minimal, including only animal care and healthcare.

Animal Care

Shelter Bedding

Wood shavings can come in bales of 8 cu. ft. Each such bale costs roughly $8. Two bales are needed to cover the floor of a 96-sq. ft. coop, 2 inches deep. The bedding should be replaced every 2–4 weeks. If the bedding is replaced every 2 weeks, the weekly cost would be $8. One straw bale will cost roughly $5 and will last for about 1 week as an added layer on the floor.

Vet Care

Vet care can vary, but the prevailing cost estimate is about $52 per year for a dozen geese.

Item	Weekly Cost
Shelter bedding (straw bale)	$5.00
Shelter bedding (wood shavings)	$8.00
Vet care	$1.00
Total	**$14.00**

Cost/Benefit

The purpose for this analysis is to provide a way to examine the cost versus the benefit of raising geese for household meat. The analysis will compare the cost of meat production to the value of comparable goose meat prices.

Savings/Loss

The cost analysis for chickens and ducks did not include the initial investment nor the expenses to get the birds to egg-producing ages. The cost analysis was based on an operating cost for weekly egg production.

Geese, on the other hand being raised for meat, will need to be replaced each year. Therefore, the initial investment and expenses for brooder feeding will be included in the seasonal cost analysis for each batch of meat birds.

The following data set is based on 12 Embden geese. The geese are fed for the first 7 weeks, then allowed to graze for an additional 9 weeks before being processed at 16 weeks of age.

Processed weight at 16 weeks = 12 lb. x 12 geese = 144 lb.

- 144 lb. x $9.50 per lb. = $1368 value

Item	Weekly Expense	Weekly Value or Revenue	Weekly Savings/ Loss
Start-up investment	$239.40		
Feeding to 8 wks.	$197.76		
Grazing for remaining 8 wks.	$0.00		
Animal care (8 wks.)	$112.00		
Total cost for 12 meat geese	**$549.16**		
Total value for 12 meat birds processed at 16 weeks		**$1368**	
Savings/loss			**+ $818.84**

The reason for the high savings figure for raising geese for meat is due to their ability to offset typical commercial feed cost by grazing. As you can see, this is a tremendous benefit.

Picture Gallery

Meat Geese

Embden

Toulouse

American Buff

Dual Purpose

Chinese

African

CHAPTER 4
Turkeys

There are numerous domestic turkey breeds that come in a variety of colors and sizes. Given that turkeys can be colorful, some homesteaders enjoy having a few around to enhance the ambiance of their homestead. For that purpose, there are several breeds that have colorful plumage and possess friendly temperaments.

Most homesteaders, however, consider raising turkeys for one objective, and that is meat. When deciding which turkey breed would be best for meat on your homestead, you basically have two varieties to choose from: Broad Breasted and Heritage.

For those wanting to raise turkeys for meat quickly and efficiently, the Broad Breasted variety is the best choice. For those not so concerned about the amount of meat, nor the time involved in production, but perhaps interested in keeping the turkeys year-round, the Heritage variety may be a better choice.

This chapter will discuss the various turkey breeds with these two varieties in mind, and how they all compare to one another regarding characteristics and meat efficiency. The typical aspects of brooding, feeding, foraging, watering, and meat production will be discussed. In addition, information will be provided regarding space requirements for shelter and containment.

The chapter will conclude with a cost/benefit analysis to examine meat production for profit and will also consider the value of household meat by comparing production costs to typical retail meat prices.

Breeds

This section divides the breeds into two categories: Broad Breasted and Heritage breeds. For those interested in raising turkeys for meat with the best feed conversion ratio, the Broad Breasted breeds are the best option. For those interested in keeping the turkeys around for a longer period of time, with less feed efficiency, but have the option for breeding, the Heritage is the way to go.

The characteristics of each breed, such as origin, plumage, adult weights, and temperament, will be discussed. In addition, there is a summary of the description of each variety with a few pros and cons. The section will conclude with a chart comparing processing weights and a few other comparison criteria.

Broad Breasted Breeds

There are two Broad Breasted breeds that are highly used by commercial growers because of their fast growth and high processing weights. The two most common breeds are the Broad Breasted White and the Broad Breasted Bronze. Both have efficient feed conversion rates, contributing to their high usage for commercial meat production.

Both of these Broad Breasted breeds are hybrids, meaning that the breed is a result of crossbreeding. Being a hybrid, they typically don't reproduce on their own and must be bred with the use of artificial insemination. One must, therefore, restock each year by purchasing poults from hatcheries.

Broad Breasted White

The Broad Breasted White breed was developed in the United States by crossing the Broad Breasted Bronze with the White Holland breed. Their plumage is made up of all white feathers. A mature adult weight typically reaches 30–45 pounds. Their temperament is docile and is well-suited for commercial production. The breed is known for its rapid growth, and the meat is widely popular for the large carcass size. The large breast is rich in white meat.

Broad Breasted Bronze

The Broad Breasted Bronze was also developed in the United States. Their plumage has dark feathers with a metallic sheen, resembling the US wild turkey. The adult live weight is similar to the Broad Breasted White, ranging from 30 to 45 pounds. Their temperament is calm and adaptable. Like the Broad Breasted White, the Bronze variety is sought after for its rapid growth rate and resulting high processing weight. The meat is considered to be popular and valued for its quality and appearance.

Heritage Breeds

Heritage breeds are numerous with a large variety of colors and sizes. Some of the Heritage breeds are used as ornamental farm pets and some are used for eggs (although they lay about a third the number of eggs as other poultry), but mostly Heritage turkeys are raised for meat.

The rate of growth and ending carcass weights are not as impressive as the Broad Breasted varieties, but they are able to forage more than the larger varieties, thus making it possible to offset some of the extra feed costs. The Heritage birds are also able to reproduce naturally, making it possible to raise replacements by hatching eggs, if that is desired.

Standard Bronze

The Standard Bronze originated from a cross between domestic English turkeys and native American wild turkeys in the United States. Their plumage is similar to the Broad Breasted Bronze, in that it resembles the American wild turkeys with dark feathers and a metallic sheen. The adult live weight is about 25–35 pounds, which is historically larger than most of the other Heritage breeds.

Chocolate

The Chocolate breed has been included among the Heritage breeds due to their popularity among southern US states during the Civil War. As their name indicates, their plumage is chocolate brown in color. Chocolates are not subject to health and mobility problems, like some of the other large Heritage breeds. The adult live weights range from 18 to 30 pounds. Their meat quality is considered to be good and flavorful. They have a gentle nature and are good foragers.

Bourbon Red

This breed was developed in the US in Pennsylvania and Kentucky. Their plumage is chestnut red with white stripes. The adult live weight is about 16–25 pounds. Their temperament is gentle, and they are well-suited for free-range systems. The breed is valued for its rich-flavored meat and is often chosen by those interested in Heritage breeds for meat.

Narragansett

The Narragansett is a cross between domestic European turkeys and native American wild turkeys. It was further developed in the Rhode Island area. The plumage is gray with black and white stripes. The adult live weight ranges from 16 to 25 pounds. They are considered to be generally calm and good foragers. Being an early Heritage breed, they are appreciated for both their meat and historical value.

White Holland

The White Holland is thought to originate from the Netherlands. Their plumage is all white feathers. The adult live weight is typically about 16–25 pounds. Their temperament is calm and friendly, making them suitable for small farms. The White Holland is known for excellent meat quality, a wide breast, and is gaining popularity for sustainable farming. The White Holland was used to create the hybrid Broad Breasted White breed.

Blue Slate

The Blue Slate breed originated from the US. The plumage has a gray-blue slate color. The adult live weights are 14–23 pounds. They have a calm, friendly temperament and are easy to manage. Their meat is considered to be tender, succulent, and flavorful. They are known to be hardy and good foragers.

Black Spanish

The Black Spanish breed originated from Europe and was introduced into the US during the colonial period. The plumage has a stunning black appearance with a greenish to bluish tint. This is a medium-sized breed, and the adult live weights range from 14 to 23 pounds. The meat quality is considered to be popular for its rich, full flavor. They are a hardy breed, capable of withstanding a variety of weather conditions.

Midget White

The Midget White breed is one of the smaller Heritage breeds and is a cross between the Royal Palm and the Broad Breasted White. The breed was developed at the University of Massachusetts in the 1950s. Their plumage is all white feathers. The adult live weights are on the smaller end, ranging from 12 to 20 pounds. They are a calm bird and are sometimes raised as pets. The meat quality is excellent and is known for being tender and flavorful. Being a smaller bird, the Midget White is a good choice for homesteads that have limited space.

Royal Palm

Royal Palm turkeys are a smaller breed and are known to originate from the US. The breed is a result of crossbreeding wild turkeys, Black, Bronze, and Narragansett. Their plumage is white in color with black outer edges. The adult live weights are 10–16 pounds. They have a docile and friendly temperament. Being smaller in size, they are often raised for both ornamental and meat purposes. The meat is considered to be tasty and flavorful. Royal Palms are hardy and can tolerate a range of climates. They are good foragers, making them a favorite for small homesteads.

Comparison Criteria

Chart 1

Criteria	BB White & Bronze	Standard Bronze	Chocolate	Bourbon Red	Narrag-ansett
Adult size	X Large	Large	Large	Med–Large	Medium
Market wt. (lb.) tom/hen	45/30	35/25	30/18	25/16	25/16
Processed wt. (lb.)	34/22	26/19	22/13	19/12	19/12
Meat quality	Good	Excellent	High	Excellent	Very Good
Reproductive success	AI	Medium	Medium	Medium—High	Medium—High
Hardiness	Medium	Good	Good	Good	Good
Temperature tolerant	Fair	Medium	Medium	Medium	Medium
Disease resistance	Medium	Medium	Medium	Medium	Medium
Foraging	Poor	Medium	Good	Good	Good

Chart 2

Criteria	White Holland	Blue Slate	Black Spanish	White Midget	Royal Palm
Adult size	Large	Medium	Medium	Small—Medium	Small—Medium
Market wt. (lb.) tom/hen	25/16	23/14	23/14	20/12	16/10
Processed wt. (lb.)	19/12	17/10	17/10	15/9	12/7
Meat quality	Excellent	Good	Good	Very Good	Very Good
Reproductive success	Low—Medium	Medium	Medium	Medium—High	Medium—High
Hardiness	Medium	Good	Good	Good	Good
Temperature tolerant	Fair	Medium	Medium	Medium	Medium
Disease resistance	Medium	Medium	Medium	Medium	Medium
Foraging	Fair	Good	Good	Good	Good

Hatching and Brooding

The incubation and brooding process is similar to that of chickens. For a detailed description of the process and various equipment options, refer to Starting with Eggs and Starting with Chicks in Chapter 1, Part 2.

Incubating Eggs

The main difference for turkey eggs is the number of days required for hatching and temperature and humidity settings. Turkey eggs require, on average, 28 days to hatch. Turkey eggs should be incubated at a temperature of 99.5 degrees F.

Humidity should be 50–60% for the first 25 days and the eggs turned five times per day. On day 26, egg turning should stop, and the humidity should be increased to 65–75% until pipping starts (day 28).

Brooders

The most common practice for starting your turkey flock is to purchase poults from a hatchery. The process and equipment for brooding turkey poults are similar to that for chicks. The main differences are that temperature and space requirements need to be more carefully observed and followed.

Poults are not as hardy as chicks, ducklings, or goslings, thus the need to stay within the heat and space parameters. Poults will need to remain in the brooder for 6–8 weeks, until fully feathered.

Brooder Space

The space requirement per poults is about 1 sq. ft. per poult for the first week. For weeks 2–3, each poult will require about 2.5 sq. ft. For weeks 4–8, each poult will require 3–4 sq. ft.

Brooder Heat

The temperature should begin at 95 degrees and be reduced 5 degrees per subsequent week thereafter until 70 degrees is reached. As a heat source, you can use either a heat plate or heat lamps. Daylight hours management isn't as essential as it is for chicks, because the poults are being raised primarily for meat instead of eggs.

If heat plates are not large enough, you can hang heat lamps above the brooder. The height would be governed by the temperature desired in the brooder. If the heat is not enough, the poults will be crowding under the heat source. In that case, the lamps need to be lowered. If the heat is too warm, the poults will be congregating around the perimeter of the brooder, and thus require the lamps to be raised accordingly.

Brooder Waterers

Self-filling poultry waterers have a saucer-shaped base, with a screw-on plastic water jar. The waterer should be positioned to allow free access on all sides, and you should have one on each end of the brooder.

Poults may need to be taught how to use the waterers. One method is to dip their beaks in the water. Once they get the idea, others will follow. Another method is to have a few chicks in the brooder with the poults. Once the poults see the chicks using the waterers and feeders, they will begin to catch on.

Because the water can become contaminated with debris, the waterer should be replenished with fresh water each day. After the first couple of weeks, the waterers should be elevated a bit to accommodate the poults' growth.

Brooder Bedding

The best bedding material for poults in the brooder is chopped straw. Wood shavings can also be used, but straw is better if available. The depth should be about 4 inches and the bedding should be changed once per week.

Brooder Tops

The top of the brooder should be covered to protect the poults from predators. Typically, the top would consist of a metal frame covered with hardware cloth or chicken wire. The most common predators for brooders are cats.

Brooder Feeding

During the first 2 weeks, each poult will consume about .04 pounds of feed per day. Between weeks 1–7, the feed consumption per poults will gradually increase from .04 lb. to .2 lb. of feed per day for Heritage breeds and .36 lb. per day for Broad Breasted breeds. During their stay in the brooder, the poults should have continuous access to feed and water.

The same round bulk type of feeder that are used for chicks will work for poults in the beginning. After about 3–4 weeks, you may need to switch to a metal trough type feeder, so that the poults can have full access to feed at all times as they grow. Feeder selections should be a free choice feeder allowing continual access. You should have one on each end of the brooder—one feeder and one waterer on each end with the heat source in the middle.

As a feed choice, poults should be fed starter crumbles, which are specially formulated for their growth and development. Turkey poults need more protein than adults, especially in the first 4 weeks of their lives. Starter feed has a protein content of 28–30%, which is ideal for the first 4 weeks. For weeks 5–8, you can reduce protein content to roughly 26%. Turkey starter feed or gamebird starter can be used for weeks 1–8, in crumble form, with high vitamin C content for stress.

Flock Feeding

Feeding your flock (rafter) is relatively easy with free choice, self-feeding equipment, but choosing the right type of feed and knowing when to switch to different rations are important.

Equipment

Turkey feeders can be the same as chicken feeders. One common feeder type has a large saucer at the bottom with a screw-on plastic feed container. The other common type is the long trough with a triangular top with large holes in the top for easy access.

Unfortunately, these types introduce several problems: waste from spillage, spoilage from wet weather, and rodent attraction. To alleviate these problems, many creative feeder alternatives are available online—most of which include a hanging version with a cone-shaped top. The key is to have a viable source of feed readily available to the turkeys at all times during the day.

Commercial Feeds

The most commercial turkey feeds include starter feed, meat-bird feed, and maintenance feed. Feeding turkeys for meat production requires specific types of feed that are formulated to supply their nutritional needs, especially protein, as they grow and develop.

Starter Feed (Weeks 1–4)

Newly hatched poults should be fed a high-protein starter feed containing about 28–30% protein for the first 4 weeks. This helps promote early growth and feather development.

You can use turkey starter at 28% protein or gamebird starter at 30%. The high-percentage protein for turkeys is what is important—chicken feed will not work. You should be able to find these at a local feed store or online.

Starter Feed (Weeks 5–8)

For weeks 5–8, the poults should be transitioned to a starter feed with slightly less protein—26%. This helps promote strong bones for rapid growth and feather development.

Meat-Bird Starter to Finish Feed (Weeks 9–16)

After the poults leave the brooder, you will need a feed to accommodate their nutritional needs as they begin to transition from growth to meat production. The meat-bird starter feed is reduced to 18–22% protein and is fortified with vitamins and minerals to meet increased energy levels and improve meat production efficiency.

Maintenance Feed (Weeks 17–28)

Turkey maintenance feed is typically 12–16% protein. It is formulated to help maintain proper muscle tone and skeletal growth and promotes meat production without excessive weight gain. It supplies a balanced nutrition for mature birds and contains prebiotics and probiotics to support immune and digestive health.

Supplemental Grains

Commercially formulated feeds are designed to meet the specific nutritional requirements of turkeys at different life stages. These feeds typically contain a mix of grains, protein sources, vitamins, and minerals to ensure the turkeys receive a balanced diet.

If you're considering adding supplemental grains, like corn and wheat to the diet, it's important to do so in moderation, and with consideration for the flock's overall nutritional balance. Too much reliance on grains alone may lead to imbalances in essential nutrients.

A common practice is to use grains as a supplement to the commercial feed rather than a primary source of nutrition. The recommended percentage of supplemental grains can vary, but a general guideline might be to keep it about 10–20% of the total diet.

Scratch Feed

As a fun addition to your formulated feed, scratch grains can be purchased to encourage natural pecking, foraging, and feeding instincts. It generally consists of whole grains, sunflower seeds, legumes, millet, and other natural ingredients. The total protein content is roughly 8%.

Because scratch feed is lower in nutrition, it should be fed in the afternoon after the turkeys have had their nutritional needs met with a complete ration feed. Then only provide enough scratch grain that can be finished in 15–20 minutes.

Foraging

It is not as easy to raise turkeys in small, confined areas, like you can with chickens, because of their size. If you plan on raising more than just a few turkeys, it would be good to have a grassy area for foraging.

While foraging can be a natural behavior for turkeys and can contribute to their overall well-being, the majority of their diet is often supplied through a balanced commercial feed. This ensures that they receive the necessary nutrients in the right proportions.

For the most part, foraging will consist of grass and weed seed, clover and other legume seed heads, and various types of insects. The percentage of the diet that comes from foraging may vary, but it's common for it to be a relatively small part of their overall nutritional intake.

Flock Watering

Water consumption is one of the most important factors in turkey performance and health. What affects water consumption the most are water quality and water availability. To optimize water quality and availability, you should consider your water management and water equipment.

Daily Water Intake

Water consumption is important for digestion, waste elimination, and body temperature. Optimum water intake will also help to optimize feed intake, which is necessary for maximum growth and meat production.

Daily water intake will begin at roughly .1 quart per day per poult, during the first week, and will increase to .5 quarts per day per poult by the time they are ready to leave the brooder at 8 weeks. Daily intake will slowly increase each week, to roughly 1 quart per day per turkey at age 20 weeks.

Water temperature and water cleanliness are important. They can be managed with proper water equipment and maintenance.

Water Equipment

Cool, clean water should be freely accessible to your turkeys at all times. Water helps to control body temperature and is important during hot weather.

Daily water intake volume will increase with age, so the size and number of waterers available will be an important consideration. On average, a mature turkey will drink about 1 quart of water per day, and perhaps 1.5–2 quarts when it is hot. A 5-gallon waterer will service 10 turkeys for a day. It is, therefore, recommended to have at least one 5-gallon waterer, on average, for every 10 turkeys.

If possible, waterers should be placed in the shade. If your turkeys are allowed to forage at all, you may want to have multiple water sources for better flock dispersal.

The most common types of waterers for turkeys are the gravity flow, self-filling type; the automatic type; the water cup type; and the nipple type. When making your waterer choices, you should consider which option best performs the following.

1. Provides cool water.
2. Promotes clean water.
3. Prevents algae formation.
4. Allows for heating to prevent freezing.

Gravity Flow Waterers—Galvanized

The galvanized gravity flow waterers consist of a container and a corresponding round trough attached to the bottom of the container. The inner tank is filled with water from the top, and an outer shell fits over the inner tank and is locked into place with a turning motion. Water is then allowed to fill the trough, and the trough water level is vacuum controlled.

Most galvanized gravity flow waterers can sit on an elevated platform or be hung. It is recommended to arrange for the height of the water source to be equal to the height of the turkey's back—roughly 12–14 inches.

Being double walled, the galvanized waterer is able to keep the water relatively cool in the summer and free of algae. For wintertime freeze protection, the galvanized waterer can be placed on an electric heated base that is thermostatically controlled.

Gravity Flow Waterers—Plastic

The plastic gravity flow waterers also consist of a container and a corresponding round trough attached to the bottom of the container. The round trough is screwed onto the container or attached by a turning lock. Trough water is vacuum controlled in some waterers and float controlled in others.

Some plastic versions are filled with water by first flipping the container upside down and unscrewing or unlocking the trough. After the container is filled with water, the trough base is screwed or relocked into position and then flipped right side up. Other plastic versions have a screw top that will allow filling from the top without the flipping.

Some gravity flow waterers are designed to sit on the ground, and some can be suspended by a hanging handle. Because trough water can become dirty if close to the ground, it is recommended to either elevate the trough level to about 12 inches off the ground by placing it on a platform or by hanging the waterer to this height off the ground.

The water level is visible but can at the same time allow water temperature to be affected by the sun in the summer and allow algae to grow. For wintertime freeze protection, some plastic versions are electrically heated.

This type of waterer works well in the brooder. Once the poults leave the brooder, you may want to switch to a galvanized version, which better accommodates larger sizes, or switch to an automatic watering system.

Automatic Waterers

True automatic waterers are those that are connected to a water source and therefore do not require refilling a container or reservoir. They can come in a variety of designs, such as a trough, water cup, or nipple. Most trough type delivery systems have a float valve for water level control. The cup type has a small flap, and the nipple type has a small toggle that controls water on demand.

Cup Waterers

This type of waterer option can be purchased with a ready-to-use container, or the cups may be purchased individually and added to a container of your choice. The cup has a small float flap, that when pushed by the turkey's beak will allow water to flow into the cup. This option is great for providing clean water on demand.

Nipple Waterers

The nipple waterer generally comes as a medium-to-large container, equipped with one or more nipples on each side of the container. The nipple is a small toggle that when pressed in or sideways allows a small flow of water on demand. This system may require a certain amount of learning time for your turkeys to fully adapt.

Meat Production

Raising turkeys for meat is a common strategy, but before proceeding with a plan, a few things should be considered, such as numbers, processing ages, and plans for the meat (selling vs. home use).

Best Processing Ages

Processing age is largely governed by the breed of turkey you have chosen to raise, which, in turn, is governed by your meat production goals. If you are wanting to raise turkeys for meat and want to have the most meat produced in the least amount of time, then the Broad Breasted breeds are the best option.

If you are wanting to raise turkeys for meat but are not as concerned about the amount or timing of meat produced, then the Heritage breeds might be a better fit. The Broad Breasted breeds will be ready to butcher in 16–20 weeks and the Heritage breeds will take 24–28 weeks to finish.

Production Performance

Meat management is all about achieving the most meat in a given time for the least amount of investment. The following figures are based on free access feeding only.

These figures can be offset with the use of foraging but will affect either timing or meat weight. If you want to using foraging, and are wanting the weights indicated below, it may take you longer to achieve. If you want to use foraging, and still process your turkeys by the typical week milestones, you may not be able to achieve the weights indicated below.

With that said, if your goal is to achieve the most meat within the typical week milestones below, you can expect to have similar meat production weights and costs provided in these charts by feeding your turkeys with minimal foraging. The figures below include the cost of feed only. They do not include poult purchase prices nor processing fees.

Broad Breasted Breed Costs

Broad Breasted turkeys are typically raised to 16 and 20 weeks for meat production.

Conversion Performance

Criteria	Total Feed Lb.	Live Weight Lb.	Feed-to-Gain Ratio (FCR)
16 weeks			
Tom	64 lb.	25 lb.	2.6
Hen	46 lb.	19 lb.	2.4
20 weeks			
Tom	101 lb.	36 lb.	2.8
Hen	64 lb.	24 lb.	2.7

Heritage Breed Costs

The Heritage breeds are typically raised to 24 and 28 weeks for meat production.

Conversion Performance

Criteria	Total Feed Lb.	Live Weight Lb.	Feed-to-Gain Ratio (FCR)
24 weeks			
Tom	56 lb.	14 lb.	4.0
Hen	40 lb.	10 lb.	4.0
28 weeks			
Tom	100 lb.	20 lb.	5.0
Hen	75 lb.	15 lb.	5.0

Please note that the feed consumption is similar for the Broad Breasted turkeys at 20 weeks and the Heritage breeds at 28 weeks. The Heritage breeds will eat less per day but for more days. The main difference will be with the processed weights.

Meat Marketing

Before you decide to raise turkeys for meat to sell, you will need to verify that you have a market for your turkey meat. There are several places where you might be able to market your meat. For a complete list of viable options for marketing, please refer to Meat Marketing (for Geese) in Chapter 3, Part 2.

Meat Processing

If you choose to butcher your own turkeys, there are a few meat processing procedure options you can consider. Please refer to Meat Processing (for Geese) in Chapter 3, Part 2 for a thorough explanation, regarding dispatching, plucking, butchering, and storage.

Shelter

In mild climate areas, a three-sided shed can be used as a shelter, but in areas that have wet weather, cold winter weather, and/or lots of predators, an enclosed shelter should be used.

Your turkeys should have a structure available that provides a safe area to roost and have protection from harsh weather and predators. If enclosed, the shelter should include windows for natural lighting and ventilation and a floor covered with litter material for moisture absorption.

Space Requirements

On average, turkeys need about 8–10 square feet each inside the shelter in order to accommodate the larger Broad Breasted variety. The indoor space should be connected to adequate outdoor space for feeding and/or exercise.

Nesting

Turkeys will lay eggs, but their egg laying is seasonal (April–June). Even though they can lay up to 100 eggs per year, hens generally lay and sit on one clutch at a time. For that reason, most homesteaders don't raise turkeys for eggs. In the spring, turkey hens will make a nest out of the straw on the floor of the shelter and will lay their eggs there. No special accommodation need be taken for egg laying other than the required straw on the floor of the shelter for moisture absorption.

Bedding Material

The floor will need to be covered with bedding material. The bedding material will serve mostly for absorbing moisture.

Generally, straw is the best bedding material for the indoor floor. The bedding material will need to be replaced every 2 weeks or so.

Roosting

Typically, turkeys like to roost on something elevated above the ground for sleeping at night. The best perches for roosting are 2" x 4"s that are arranged horizontally to the floor inside the coop.

An easy design is to cut two 2" x 4"s, 4 ft. long, which will be used to support the perches. Determine where you want your roosting bars or perches to be situated, and place the 4-ft. supports on each end, about 4 ft. apart. Arrange for the top of the support to be attached to the wall, about 36 inches above the floor, and then angle out at roughly 45 degrees, so that the bottom of the support is attached to the floor, roughly 32 inches from the wall.

Next, cut two notches in each support (like a staircase), on which the two perches are to be placed. The bottom perch should be placed roughly 10 inches above the floor and 24 inches from the wall. The upper perch can be situated roughly 25 inches above the floor and 10 inches from the wall. Cut two perches, 6 feet long, and place them into the notches in the supports and attach them to the end supports. You will have two perches, 6 feet long, that will extend 1 foot beyond the supports on each end. This roosting arrangement will allow each hen to have 15 inches of roosting space.

The turkeys will poop the most while roosting, and, therefore, a lot of poop can build up under the roosting perches. A latched, trap door can be built into the floor under the perches. Such an arrangement allows for easy access when the floor needs to be cleaned. Just open the trap door and allow the soiled litter to fall to the ground for easy removal. Then relatch the door closed and add fresh litter inside.

Construction

There are many ways to construct a turkey coop, and you can find a variety of plans online. You can also find many prefabricated kit versions online. The most common plans are those that use wood for construction material.

In the Chicken Coops section in Chapter 1, Part 2, instructions were given on how to build a chicken coop for an example flock of 12 chickens. That same type of structure can be used for an example flock of 10 turkeys, but the structure size would need to be double—a total of 96 square feet in order to meet the minimum square feet per turkey requirement.

For this example, the dimensions and building material can be doubled from the example in Chapter 1, Part 2. The only difference is that nesting boxes would not be needed, and roosting bars will require different spacing.

Like the chicken coop, the structure should be elevated to better protect the turkeys from predators, from flooding when it rains, and for facilitating soiled litter removal.

Floor

The best plans for a coop are those that are elevated off the ground. The advantage of having the floor elevated is to reduce the threat of rodents and snakes and to reduce moisture problems when it rains.

To be elevated, it is best to begin by installing 6-inch diameter round posts or 4" x 4" treated timber posts in the ground at all four corners, plus additional center supports in front and back. The posts should be secured in the ground with concrete and should extend about 14–16 inches above the ground. To accommodate a 96-sq. ft. structure for turkeys, the corner posts should be 8 feet apart on the sides, and 12 feet on the front and back (with one support post in the middle of the 12 ft. span in front and back).

The frame for the floor would be constructed with 2" x 6" boards, and rest on top of the support posts. The frame would consist of the outer frame and inner floor joists. Then 4' x 8' sheets of ¾" plywood can be cut and applied to the top of the frame for the floor.

Walls

For this example, the front wall will be 6 ft. tall and the back wall 5 ft. tall. The walls can be framed with 2" x 4"s, and 4' x 8' sheets of ½" plywood can be cut and applied to the outer frame for the walls. Windows can be cut through the walls for ventilation. The windows are typically above the roost in the front and high in the back for cross ventilation. The windows should be covered with screen or hardware cloth to keep out rodents and predators and should have shutters that can be closed at night and during harsh weather.

Roof

The roof can be arranged to be gabled in the front and back with the pitch slanting toward each side, or you can have the front wall higher than the back wall so that the roof slants toward the back. There are a variety of methods and models to choose from. For this example, simply arrange for the front wall to be about 1 ft. higher than the back wall, allowing the roof to slant toward the back with a 2–12 pitch.

Doors

There are many options to choose from, but most prefer to have a small door at the front corner with a ramp to the outside turkey pen for the turkeys to have easy access in and out. In addition, a separate, larger door would allow access for the homesteader to enter for cleaning and maintenance.

The doors should have latches that can be closed during the night. The latches should be secure and lockable.

Ramp

If the coop is elevated, you will need to build a ramp so that the turkeys can have easy access in and out. The ramp can be made of wood. The ramp should be built in such a way so that it doesn't sag in the middle, and should have 1" x 2" cross boards on the surface to prevent slipping when wet. The incline of the ramp should be shallow to accommodate large turkeys.

Electricity

Turkeys are raised primarily for meat, so it is not mandatory to have electricity installed in the shelter for electric lighting. The lighting is a requirement for poultry that are being raised for eggs, which need 15–17 lighted hours each day.

You may need electricity for heat in the winter, but generally, turkeys are hardy and can withstand cold weather fairly well.

Feeding and Watering

Feed and water should be kept outside whenever possible. You may need heated waterers to prevent freezing in the winter.

Containment

In addition to the shelter, turkeys should also have an outside pen for containment. This is mainly for protection from predators and when they are not able to forage when the weather is bad.

Space Requirement

The minimum required space per turkey in an outside enclosure is 20 square feet per turkey. For an example flock of 10 turkeys, the outside square footage would need to be a minimum of 200 square feet. An outside pen that is roughly 30 ft. x 8 ft. producing 240 square feet would suffice.

In the Chicken Run section in Chapter 1, Part 2, a description was given for an outside pen that was 16 ft. x 8 ft. The same method of construction and type of material can be used—it would just need to be twice as long.

Unique Behavior

Turkeys are relatively easy to raise without any major behavior problems. You can generally allow toms to run with the hens without any difficulty of aggressiveness.

The only potential problem is that turkeys can tend to be a bit flighty. They can take flight more readily than other poultry breeds and can fly over fences, roost in trees, etc. They will generally stay close to home, if there is plenty of food and water available.

You will really need to take care and house the turkeys at night, as predators will attack the turkeys for food.

Healthcare

Raising turkeys can be a rewarding experience, but they can be susceptible to various health issues, like any other poultry on a homestead. Listed below are some typical health issues that a small flock of turkeys on a homestead might encounter.

Respiratory Infections

Turkeys are prone to respiratory infections, often caused by bacteria or viruses. Common symptoms include coughing, sneezing, and nasal discharge.

Parasites

Internal parasites, such as worms, and external parasites, like mites and lice, can affect turkeys. Regular deworming and proper sanitation can help prevent parasite infestations.

Coccidiosis

This is a common protozoan infection that affects the digestive tract of turkeys. It can lead to diarrhea, weight loss, and lethargy. Good cleanliness practices and proper replacement of bedding material can help prevent coccidiosis.

Blackhead Disease (Histomoniasis)

Caused by a protozoan parasite, blackhead disease primarily affects turkeys. It can result in liver damage and respiratory distress. The best preventive measure includes maintaining a clean environment and by not allowing your turkeys to share the same coop with chickens, as chickens can be carriers.

Heat Stress

Turkeys are sensitive to temperature extremes, especially heat. Heat stress can lead to respiratory problems. Providing adequate shade, ventilation, and fresh water are important in preventing heat-related issues.

Botulism

Contaminated feed or water sources can lead to *botulism* in turkeys. This toxin-producing bacteria can cause paralysis and death. Proper sanitation and monitoring of feed and water quality are essential preventive measures.

By regularly monitoring your flock health, vaccinating when applicable, and pursuing veterinary assistance when warranted, you can manage and/or prevent many of these health issues in a small homestead flock of turkeys.

Expenses

The following expense data will compare figures for Broad Breasted and Heritage turkeys, beginning by purchasing 10 poults from a hatchery.

Start-Up Expense Summary Charts

Start-up expenses will include the cost of equipment and housing. Purchase cost for poults will be included in the operating cost, as turkeys will be used for meat and therefore need to be replaced each season.

Infrastructure and Equipment Expense Chart

Item	Cost
Brooder box	$45
Heat lamp for light	$18
Brooder heat plate	$45
Poult feeder & waterer set	$18
Poult trough feeder	$23
Flock feeders	$25
Flock waterers	$35
Turkey shelter— ($13.50/sq. ft.)	$1296
Turkey pen— ($2.00/sq. ft.)	$480
Total	**$ 1985**

Operating Expenses

Given that turkeys will be raised for meat, the flock will be replaced each year. The operating expenses will, therefore, include the purchase price for the poults. Operating expenses will also include the feed cost from week 1 to weeks 16–20 for Broad Breasted breeds and up through weeks 24–28 for the Heritage breeds. Animal care will also be included in the weekly operating expenses.

Poults Purchase Costs

Item	Cost
Price per poult for Broad Breasted	$9.00
Price per poult for Heritage	$17.00

Feed Pricing Chart

The following feed pricing will provide the standard to be used for both the Broad Breasted and Heritage turkey weekly feed costs.

Weeks	Feed	Cost per Bag	Cost per Lb.
Weeks 1–4	Game bird starter-30%	$26/40 lb.	$.65/lb.
Weeks 5–8	Turkey starter-25%	$34/50 lb.	$.68/lb.
Weeks 9–16	Meat bird-22%	$24/50 lb.	$.48/lb.
Weeks 17–28	Maintenance-16%	$24/50 lb.	$.48/lb.

Broad Breasted Feed Costs (per Poult)

Weeks	Feed Amount (Lb.)	Feed Cost	Total Feed Cost at 16 & 20 Wks.
Weeks 1–4		$.65/lb.	
Tom	2.75	$1.79	
Hen	2.3	$1.49	
Weeks 5–8		$.68/lb.	
Tom	10.05	$6.83	
Hen	8.27	$5.62	
Weeks 9–16		$.48/lb.	
Tom	51.28	$24.61	**$33.23**
Hen	35.2	$16.90	**$24.01**
Weeks 17–20		$.48/lb.	
Tom	41.64	$19.99	**$53.22**
Hen	28.6	$13.73	**$37.74**

Heritage Feed Costs (per Poult)

Weeks	Feed Amount (Lb.)	Feed Cost	Total Feed Cost at 24 & 28 Wks.
Weeks 1–4		**$.65/lb.**	
Tom	2.6	$1.69	
Hen	2.3	$1.50	
Weeks 5–8		**$.68/lb.**	
Tom	5.36	$3.65	
Hen	4.74	$3.22	
Weeks 9–16		**$.48/lb.**	
Tom	26.56	$12.75	
Hen	19.02	$9.13	
Weeks 17–24		**$.48/lb.**	
Tom	42.07	$20.19	**$38.28**
Hen	29.72	$14.27	**$28.12**
Weeks 25–28		**$.48/lb.**	
Tom	23.75	$11.40	**$49.68**
Hen	17.08	$8.20	**$36.32**

Animal Care

Shelter Bedding

Straw is the best bedding for turkeys. One bale can be used to cover the 96-sq.-ft. floor at 2–4 inches deep and should be replaced every 2 weeks or so. One straw bale will cost roughly $5. If replaced every 2 weeks, the weekly cost would be $2.50.

Vet Care

Vet care can vary, but the prevailing cost estimate is ~$52 per year for 10 turkeys.

Animal Care Weekly Costs

Item	Weekly Cost
Shelter bedding (straw bale)	$2.50
Vet care	$1.00
Total	**$3.50**

Cost Analysis

The purpose of this analysis is to provide a way to examine the total operating cost for Broad Breasted meat production and Heritage meat production, for the example flock of 10 turkeys. These figures will be used for the cost/benefit analysis.

Production Cost for Broad Breasted

Total cost at 16 weeks (flock of 10)

Item	Total Cost Toms	Total Cost Hens
Purchase price for poults	$90.00	$90.00
Feed cost @16 wks.	$332.30	$240.10
Animal care	$56.00	$56.00
Total operating cost	**$478.30**	**$386.10**

Total cost at 20 weeks (flock of 10)

Item	Total Cost Toms	Total Cost Hens
Purchase price for poults	$90.00	$90.00
Feed cost @ 20 wks.	$532.20	$377.40
Animal care	$70.00	$70.00
Total operating cost	**$692.20**	**$537.40**

Production Cost for Heritage

Total cost at 24 weeks (flock of 10)

Item	Total Cost Toms	Total Cost Hens
Purchase price for poults	$170.00	$170.00
Feed cost @24 wks.	$382.80	$281.20
Animal care	$84.00	$84.00
Total operating cost	**$636.80**	**$535.20**

Total cost at 28 weeks (flock of 10)

Item	Total Cost Toms	Total Cost Hens
Purchase price for poults	$170.00	$170.00
Feed cost @ 28 wks.	$496.80	$363.20
Animal care	$98.00	$98.00
Total operating cost	**$764.80**	**$631.20**

Cost/Benefit

The cost/benefit analysis is useful in answering the typical question if it is more cost-effective to raise your own meat or purchase it. The results of this analysis will answer that question.

Savings/Loss

By using the production figures above, you can determine the cost per pound for meat produced and the cost per processed bird. You can then compare those figures to what it would cost to purchase the meat. The results show a benefit and savings to raise your own meat, if you are willing to make the investment for infrastructure and supply the labor. The cost for infrastructure and labor is not included in these figures.

Broad Breasted Cost/Benefit Comparison

Broad breasted turkeys are typically raised to 16 and 20 weeks for meat production.

Production meat cost vs. retail cost

Weeks	Production Cost—10 Turkeys	Processed Meat Lb.— 10 Turkeys	Cost per Lb.	Retail Price per Lb.	Cost/ Processed Bird	Price/ Retail Bird
16 wks.						
Tom	$478.30	200	$2.39	$5.00	$47.80	$100.00
Hen	$386.10	150	$2.57	$5.00	$38.55	$75.00
20 wks.						
Tom	$692.20	290	$2.39	$5.00	$69.31	$145.00
Hen	$537.40	190	$2.83	$5.00	$53.77	$95.00

Heritage Cost/Benefit Comparison

Heritage turkeys are typically raised to 24 and 28 weeks for meat production.

Production meat cost vs. retail cost

Weeks	Production Cost—10 Turkeys	Processed Meat lb.— 10 Turkeys	Cost per Lb.	Retail Price per Lb.	Cost/ Processed Bird	Price/ Retail Bird
24 wks.						
Tom	$636.80	100	$6.37	$8.00	$63.70	$80.00
Hen	$535.20	80	$6.69	$8.00	$53.52	$64.00
28 wks.						
Tom	$764.80	150	$5.09	$8.00	$76.35	$120.00
Hen	$631.20	110	$5.74	$8.00	$63.14	$88.00

Results

By examining the results from the comparisons above, the 20-week Broad Breasted tom turkeys show the most monetary benefit and value—a $75 difference between processed bird and retail bird. This would yield a $750 benefit per batch of 10 turkeys.

Picture Gallery

Broad Breasted Turkeys

BB White

BB Bronze

Heritage Turkeys

Standard Bronze

Chocolate

Bourbon Red

Narragansett

White Holland

Blue Slate

Black Spanish

Midget White

Royal Palm

PART 3

Key Book Insights

Part 3 is devoted to highlighting the key takeaways from the book. These insights are basically the ultimate information that homesteaders are interested in.

For instance, these summaries include Chapter 1, Most Suitable Options for milk, eggs, meat, and fiber. Also included is a best fit scenario for a typical small homestead.

Chapter 2, Meat Highlights, provides meat conversion ratios for all of the farm animals as well as meat performance summary data.

The most sought-after information is provided in Chapter 3, Cost Per Product Unit. This information is provided in charts for both livestock and poultry.

This type of summary information is what an accountant would call the bottom line—the final analysis for production benefit and cost.

CHAPTER 1

Most Suitable Options

Milk

The decision-making factor for milk is generally the volume of milk needed, rather than the cost per gallon. Because a cow can produce so much milk per day, the cost per gallon is actually cheaper than a milk goat.

Dairy goats, however, are smaller animals and thus require smaller pens and less feed. For homesteaders wanting a low-to-moderate amount of milk, the goat is generally the preferred option.

Eggs

Even though ducks can produce more eggs than other poultry options, chickens are the easiest, cleanest, quietest, and most simple to contain and care for. Chickens are, therefore, the most viable option for eggs on small homesteads.

Meat

There are many options for meat on a small homestead. After comparing space, feed cost, and animal care, rabbits and chickens are the cheapest, simplest, and easiest option for meat for a small family on a homestead.

Revenue

Some homesteads are interested in generating revenue in addition to producing their own food needs. The most optimal option for generating revenue on a small homestead is pigs, because they can have two litters per year and can have about 10 piglets per litter.

Fiber, Wool, and Fur

Raising goats for mohair or cashmere is perhaps the only fiber product that is a primary goal. Sheep fleeces and rabbit pelts are generally a by-product of the more primary goal of meat.

Best Small Homestead Scenario

A family of four can have all of their needs for milk, eggs, and meat in a relatively small space by using goats, chickens, and rabbits. By having 1 milk goat, 6 chickens, and 4 rabbits (1 buck and 3 does), you could have, on average, 7 gallons of milk, 2 dozen eggs, and 4.8 lb. of meat per week.

CHAPTER 2

Meat Highlights

Feed Conversion Ratios

The feed conversion ratio (FCR) is a way of measuring how efficiently an animal converts feed into meat. FCR is typically expressed in terms of pounds of feed to produce 1 pound of meat.

Animal	FCR
Beef cattle	6.5–9.0 :1
Goats	4.0–6.0 :1
Sheep	4.0–6.0 :1
Pigs	3.0–4.0 :1
Rabbits	3.0–4.0 :1
Chickens	1.6–2.0 :1
Ducks	2.5–3.0 :1
Geese	2.5–3.5 :1
Turkeys	1.8–2.5 :1

For livestock, pigs and rabbits have the best FCR. This means that you can produce meat with pigs and rabbits at a relatively low cost per pound.

Poultry have a much lower FCR than livestock, meaning lower cost per pound. With chickens, you can have both eggs and meat at a lower cost.

Meat Performance

The following charts provide meat production projections at the optimum ages for livestock and poultry.

Livestock

Animal	Weeks to Optimum Sale	Meat or Auction (Each)
Calf (auction)	12–16 wks.	350 lb. auction
Kid goat (2)	12–16 wks.	30 lb. meat each
Lamb (2)	20–24 wks.	90 lb. auction each
Pig (10) x 2 litters	28 & 40 wks.	115 & 180 lb. meat ea.
Rabbit (7) x 4 litters	8–10 wks.	3 lb. meat each

Poultry

Bird	Weeks to Optimum Sale	Processed Meat	Weeks to Optimum Sale	Processed Meat
Chicken	6–8 wks.	5–6 lb. broiler	12–16 wks.	7–9 lb. roaster
Duck	6–8 wks.	4–5 lb.	N/A	N/A
Goose	16 wks.	12 lb.	20 wks.	14 lb.
Turkey	16 wks.	20 lb.	20 wks.	29 lb.

CHAPTER 3

Cost per Product Unit

Livestock

Animal	Product	End Yield	Cost per Unit Produced	Retail Cost per Unit
Milk cow (1)	Milk	1224 gal./ yr.	**$2.00/gal**	$4.30
Milk goat (1)	Milk	288 gal./yr.	**$3.78/gal**	$20.00
Goats (2 kids)	Meat	54 lb./yr.	**$3.50/lb.**	$7.00
Sheep (2 lambs)	Meat	90 lb./yr.	**$2.50/lb.**	$5.00
Pigs (hanging wt. per litter of 10) [28 wks.]	Meat	1150 lb.	**$3.02/lb.**	$4.75
Pigs (hanging wt. per litter of 10) [40 wks.]	Meat	1800 lb.	**$3.17/lb.**	$4.75
Rabbit (1) - (4 litters)	Meat	84 lb./yr.	**$3.35/lb.**	~$6.00

Poultry

Animal	Product	End Yield	Cost per Unit Produced	Retail Cost per Unit
Chickens (12)	Eggs	4 doz./wk.	**$3.60/doz.**	$3.50
Ducks (12)	Eggs	6 doz./wk.	**$5.53/doz.**	$5.50
Chickens (12) – 8 wks.	Meat	72 lb.	**$1.22/lb.**	$3.49
Geese (12) – 16 wks.	Meat	144 lb.	**$3.81/lb.**	$9.50
Turkeys BB 16 wks. Toms (10 birds)	Meat	200 lb.	**$2.39/lb.**	$5.00/lb.
Turkey BB 20 wks. Toms (10 birds)	Meat	290 lb.	**$2.39/lb.**	$5.00/lb.
Turkey Heritage 24 wks. toms (10 birds)	Meat	100 lb.	**$6.37/lb.**	$8.00/lb.
Turkey Heritage 28 wks. toms (10 birds)	Meat	150 lb.	**$5.09/lb.**	$8.00/lb.

Conclusion

By now, you can easily see how important it is to be enlightened of all the options for choosing and raising your farm animals. This book is designed to help you be informed so that you can begin with the greatest chance of success.

With the knowledge you have gained from this book, you can now plan and prepare with confidence. You now know exactly what animals you will need and which breeds will be best to accomplish your goals for milk, meat, ßeggs, and fiber.

Now that you have a plan, you can implement the plan effectively. There is no better feeling than being well prepared—knowing what to do and how to do it.

Once you have prepared for your animals by having adequate containment, shelter, and equipment, you can now properly manage the animals in an efficient manner to best achieve your goals.

With what you have learned from this book, you are well on your way toward your end goal of being self-sufficient with confidence and satisfaction.

About the Author

John Utterback grew up on a 400-acre farm in New Mexico and has enjoyed his various careers as a wildlife biologist, range conservationist, missionary pilot, airline pilot, agriculture business owner, and county agriculture standards inspector.

He enjoys compiling and organizing technical data and explaining it in an interesting and understandable manner. Technical writing is his passion; helping others is his calling.

Book Reviews

If you have enjoyed this book and found it helpful, please let me know your impressions by way of posting a book review on Amazon. It is always good to hear how people are progressing in their pursuit of their off grid dream. Thank you!

References

PART 1: LIVESTOCK
Chapter 1: Milk Cows

- Nicole Cosgrove, "5 Best Dairy Cow Breeds for Milk Production (With Pictures)," February 6, 2024, *https://www.petkeen.com/best-dairy-cow-breeds-for-milk*
- Alice, "What Are the Best Cows for the Homestead?" *https://ourdailyhomestead.com/what-are-the-best-cows-for-the-homestead/*
- Michelle and Jay Lancaster, "Comparison of the Dairy Breeds," *https://spiritedrose.wordpress.com/jersey-cattle/looking-for-a-cow/comparison-of-the-dairy-breeds/*
- Michael A. Wattiaux and W. Terry Howard, "Feeds for Dairy Cows," *https://nydairyadmin.cce.cornell.edu/uploads/doc_97.pdf*
- Donna Amaral-Phillips, "Water Intake Determines a Dairy Cow's Feed Intake and Milk Production," *https://afs.ca.uky.edu/files/water_intake_determines_a_dairy_cows_feed_intake_and_milk_production.pdf*
- Arnout Dekker and Julian Sander, "Mastitis in Dairy Cows," *https://europe.pahc.com/challenges/mastitis*
- Milky Day, "Milk Processing Equipment for Small-Scale Dairy Farm," August 24, 2020, *https://milkyday.com/blog/2020/08/24/milk-processing-equipment-for-small-scale-dairy-farm/*

- M.J. Trlica, "Grass Growth and Response to Grazing," 3/2013, *https://extension. colostate.edu/topic-areas/natural-resources/ grass-growth-and-response-to-grazing-6-108/*
- BCRC, "Two Methods, Four Steps for Calculating Carrying Capacity," March 31, 2022, *https://www.beefresearch.ca/ blog/calculating-carrying-capacity/*
- Kathy McCune, "Is Having a Family Milk Cow Affordable?" January 29, 2020, *https://familyfarmlivestock.com/ is-having-a-family-milk-cow-affordable-budget-included/*

Chapter 2: Milk Goats

- Jennifer Sartell, "Raising Dairy Goats: A Guide to Breeds," January 11, 2018, *https://www.mannapro.com/ homestead/dairygoatbreeds*
- Cheryl Howell, "Best Goat Breeds for Homesteading (13 Best Goats for Milk and Meat)," February 24, 2022, *https://www.thehappychickencoop.com/ best-goat-breeds-for-homesteading/*
- Heather Smith Thomas, "Getting Started with the Best Goats for Milk," January 3, 2022, *https://backyardgoats.iamcountryside.com/ home-dairy/a-beginners-guide-to-the-best-goats-for-milk/*
- Brian Tarr, "Guidelines to Feeding and Management of Dairy Goats," *https://epa-prgs.ornl.gov/radionuclides/ Goat_Guidelines.pdf*
- Kipp Brown, "Common Diseases in Goats," *https:// extension.msstate.edu/sites/default/files/topic-files/goats/ commongoatdiseases.pdf*
- April Lee, "How Much Does a Goat Cost?—2021 Prices" June 2, 2022, *https://farmhouseguide.com/ how-much-does-a-goat-cost/*
- Heather, "How Much Does it COST to Keep Goats? Cost Breakdown and Comparisons," 2022, *https:// sageandstonehomestead.com/*

Chapter 3: Utility Goats

Meat and Brush Goats

- Rebecca and Steve, "Brush Goats: 8 Best Tips—Brush Control & Land Clearing," April 13, 2020, *https://www.goatfarmers.com/blog/brush-goats-brush-control-land-clearing%2F*
- *David Fernandez, "Using Goats for Brush Control as a Business Strategy," http://www.uapb.edu/sites/www/uploads/SAFHS/FSA-9604.pdf*
- *Sarah Toney, "Poisonous Plants for Goats + How to Prevent and Treat Plant Toxicity, https://thefreerangelife.com/poisonous-plants-for-goats/*
- Purina Animal Nutrition, "How to Maximize Meat Goat Rate of Gain," January 8, 2021, *https://www.wisfarmer.com/story/news/2021/01/08/maximizing-rate-gain-meat-goats/6592487002/*

Fiber Goats

- Kim Irvine, "10 Best Goat Breeds for Fiber Production," March 12, 2019, *https://domesticanimalbreeds.com/the-best-goat-breeds-for-fiber-production/*
- Happy Chicken, "Fiber Goats," May 19, 2022, *https://www.thehappychickencoop.com/fiber-goats/*
- Christine Farr, "Cashmere Goats: Breeds, Characteristics, Care and Tips, December 16, 2023, *https://agronomag.com/cashmere-goats/*
- Jennifer Sartell, "Making Money with Angora Goats, Part 4," February 26, 2018, *https://homestead.motherearthnews.com/making-money-angora-goats-part-4/*
- Tara Dodrill, "Raising Fiber Goats: Everything You Need to Know," October 3, 2023, *https://www.newlifeonahomestead.com/fiber-goats-homestead-hustle/*

- Liam Maart, "Angora Goat—A Great Farm Edition," June 13, 2023, *https://www.animalsaroundtheglobe.com/angora-goat/*
- Mohair Council of America, "The Mohair Process," *https://www.mohairusa.com/pages/the-mohair-process*
- Ronan Country Fibers, "Ronan Country Fibers," *https://ronanfibers.com/wool-mohair/*
- Blackberry Ridge, "Blackberry Ridge Woolen Mill Price List for Services," July 15, 2019, *https://www.blackberry-ridge.com/pricserv.htm*

Chapter 4: Sheep

- Susan Schoenian, "A Beginner's Guide to Raising Sheep—Reproduction in the Ewe," April 19, 2021, *https://www.sheep101.info/201/ewerepro.html*
- Kathy McCune, "Where to Sell Your Wool: Options for New Sheep Owners," September 10, 2019, *https://familyfarmlivestock.com/where-to-sell-your-wool-options-for-new-sheep-owners/*
- Boyd Brady, "Common Diseases of Dairy Goats and Sheep," August 12, 2019, *https://www.aces.edu/blog/topics/sheep-goats/common-diseases-of-dairy-goats-and-sheep/*
- Steven H. Umberger, "Feeding Sheep," 2009, *https://extension.unl.edu/statewide/lincolnmcpherson/410-853_pdf.pdf*
- *USDA-NASS, "Texas Sheep and Wool," January 31, 2023, https://www.nass.usda.gov/Statistics_by_State/Texas/Publications/Current_News_Release/2023_Rls/tx-sheep-2023.pdf*
- USDA, "Public Auction Yards Sheep & Goat Auction—Billings, MT," January 22, 2024, *https://www.ams.usda.gov/mnreports/ams_1772.pdf*

Chapter 5: Pigs

- Kathy McCune, "Best Feed for Raising Pigs," June 28, 2021, *https://familyfarmlivestock.com/best-feed-for-raising-pigs-including-a-handy-chart/*
- *Brenda de Rodas, "Sow Gestation vs. Lactation Rations," https://www.purinamills.com/swine-feed/education/detail/sow-gestation-vs-lactation-rations*
- PennState Extension, "So You Want to Raise Hogs?" February 27, 2021, *https://extension.psu.edu/so-you-want-to-raise-hogs*
- Tom Guthrie, "Water Needs of Pigs," May 12, 2011, *https://www.canr.msu.edu/news/water_needs_of_pigs*
- Pork Information Gateway, "Space Requirements of Finishing Pigs Fed to a Heavier Weight (Removed Individually)," September 11, 2006, *https://porkgateway.org/resource/space-requirements-of-finishing-pigs-fed-to-a-heavier-weight-removed-individually/*
- Kathy McCune, "How Much Space Do I Need to Raise Pigs?" December 16, 2019, *https://familyfarmlivestock.com/how-much-space-do-i-need-to-raise-pigs/*
- Gene Pirelli, Dale Weber, Scott Duggan, Melissa Fery, and Nathan Parker, "Care and Management of New Feeder Pigs," December 2019, *https://catalog.extension.oregonstate.edu/sites/catalog/files/project/pdf/em9271.pdf*
- Denis Reich, James Kliebenstein, "Economics of Breeding, Gestating and Farrowing Hogs in 'Natural Pork' Production; Financial Comparison," 2006, *https://porkgateway.org/wp-content/uploads/2015/07/economics-of-breeding-gestating-and-farrowing-hogs-in-natural-pork-production-financial-comparison1.pdf*

Chapter 6: Rabbits

- Leah, Shelton, "These Are the 15 Best Rabbit Breeds for Meat," December 16, 2023, *https://agronomag.com/best-rabbit-breeds-for-meat/*
- Everbreed, "How Often Should You Breed Your Rabbits?," June 28, 2022, *https://everbreed.com/blog/how-often-should-you-breed-your-rabbits/#:~:text=Even%20though%20rabbits%20are%20biologically,length)%20is%20roughly%2030%20days*
- MSU Extension, "Rabbit Tracks: Breeding Techniques and Management," April 24, 2017, *https://www.canr.msu.edu/resources/rabbit_tracks_breeding_techniques_and_management*
- Alyssa, "Commercial Meat Rabbit Growth Rates," October 11, 2019, *https://homesteadrabbits.com/meat-rabbit-growth-rates/*
- Sarah, "Preparing for and Caring for Baby Bunnies," 2022, *https://www.livingtraditionshomestead.com/*
- Heather, "Raising Meat Rabbits: Rabbit Nutrition Basics," 2022, *https://sageandstonehomestead.com/*
- Tiffany, "Wire Flooring for Rabbits," 2023, *https://www.tealstonehomestead.com/*
- Kathy McCune, "Is Raising Your Own Meat Rabbits Worth It?" December 28, 2021, *https://familyfarmlivestock.com/is-raising-your-own-meat-rabbits-worth-it/*

PART 2: POULTRY

Chapter 1: Chickens

- Michael Schneider, "The Basics of Raising Backyard Chickens," *https://www.homestead.org/poultry/basics-of-raising-backyard-chickens/*

- Alexa Lehr, "How to Raise Chickens: 7 Essential Components," March 12, 2021, *https://grubblyfarms.com/blogs/the-flyer/how-to-raise-chickens?*
- Kathy McCune, "Is Raising Chickens for Eggs Worth It?" September 14, 2020, *https://familyfarmlivestock.com/is-raising-chickens-for-eggs-worth-it/*
- Katie Krejci, "The 7 Best Meat Chickens for Your Homestead," January 13, 2023, *https://thehomesteadingrd.com/best-meat-chickens/*
- Jacquie Jacob, "Feeding Chickens for Egg Production in Small and Backyard Flocks," *https://poultry.extension.org/articles/feeds-and-feeding-of-poultry/feeding-chickens-for-egg-production/*
- Purina Animal Nutrition, "Hatching Eggs at Home: A 21-Day Guide for Baby Chicks," *https://www.purinamills.com/chicken-feed/education/detail/hatching-eggs-at-home-a-21-day-guide-for-baby-chicks*
- Mandi Chamberlain, "Chicken Hatching with an Egg Incubator," *https://www.mannapro.com/homestead/hatching-chicks-in-an-egg-incubator#:~:text=For%20hatching%20chicken%20eggs%2C%20the,that%20but%20remember%20those%20numbers!*
- Jess, "How to Set Up a Chick Brooder," *https://mamaonthehomestead.com/how-to-set-up-a-chick-brooder/#:~:text=The%20ideal%20temperature%20for%20the,the%20brooder%20is%20draft%2Dfree*
- Dianne, "Setting Up a Brooder Box for Chicks," *https://www.hiddenspringshomestead.com/brooder-box-for-chicks/*
- The Featherbrain, "The BEST Chicken Coop Bedding: Sand vs. Straw vs. Pine Shavings," *https://www.thefeatherbrain.com/blog/best-chicken-coop-bedding*

- Manitoba Agriculture, "Basic Feeding Programs for Small Chicken Flocks," *https://www.gov.mb.ca/agriculture/ livestock/poultry/basic-feeding-programs-for-small-chicken-flocks.html*
- Germán Bertsch, "Water Quality in Poultry Production," September 27, 2019, *https://www.veterinariadigital.com/ en/articulos/water-quality-in-poultry-production/*
- Linnea, "Best Chicken Waterers: Heated & Automatic Options," *https://www.thefarmerscupboard.com/blogs/ chickens-poultry/best-chicken-waterers*
- Jill Winger, "Beginner's Guide to Raising Laying Hens," June 18, 2020, *https://www.theprairiehomestead. com/2020/03/raising-laying-hens.html*
- Chris Lesley, "Building a Safe and Sturdy DIY Chicken Coop: Step-by-Step," November 27, 2023, *https://www.almanac. com/raising-chickens-101-how-build-chicken-coop*
- RCR Organic Feed Store, "Poultry Feed Charts," *https:// www.rcrorganicfeedstore.com/Poultry-Feed-Charts*
- Amy, "Feeding Cornish Cross Chickens: How Much Should They Eat?" *https://afarmishkindoflife.com/ feeding-cornish-cross-chickens/*
- *Katie Krejci, "The 7 Best Meat Chickens for Your Homestead," January 13, 2023, https:// thehomesteadingrd.com/best-meat-chickens/*

Chapter 2: Ducks

- Jessica Knowles, "The Perfect Duck Breeds to Add to Your Homestead," July 31, 2023, *https://104homestead.com/ duck-breeds/*
- Kathy McCune, "16 Duck Breeds for Eggs and Meat" February 12, 2019, *https://familyfarmlivestock. com/16-duck-breeds-for-eggs-and-meat/*
- Wesley Hunter, "One Farmer's Guide to Raising Ducks for Meat," *https://smallfarmersjournal.com/ one-farmers-guide-to-raising-ducks-for-meat/*

- Nantahala Farm & Garden, "Incubating and Hatching Chicken, Duck & Turkey Eggs," *https://www.nantahala-farm.com/chicken-duck-incubate-eggs-s.shtml*
- Peter R. Ferket and Gary S. Davis, "Feeding Ducks," 2020, *https://poultry.ces.ncsu.edu/backyard-flocks-eggs/other-fowl/feeding-ducks/*
- Jacquie Jacob, "Feeding Ducks for Egg Production in Small Flocks," 2024, *https://poultry.extension.org/articles/feeds-and-feeding-of-poultry/feeding-ducks-for-egg-production-kept-in-small-flocks/*
- *Sharpes Farm Feeds, "Diet Requirements for Backyard Ducks—A Comprehensive Guide," https://www.stockfeed.co.nz/resources/poultry-feed/ducks-diet-requirements/#:~:text=For%20older%20ducklings%20(three%20to,for%20you%20to%20dish%20out.*

Chapter 3: Geese

- Thomas Nelson, "Comprehensive Guide to Incubating Goose Eggs," July 17, 2023, *https://thegardenmagazine.com/comprehensive-guide-to-incubating-goose-eggs/*
- Kathy McCune, "What Are the Best Geese for Eating?" April 16, 2021, *https://familyfarmlivestock.com/what-are-the-best-geese-for-eating/*
- Joel Bowers, "How to Clean and Pluck a Goose," November 10, 2020, *https://decoypro.com/how-to-clean-a-goose/*
- Jackie Linden, "Weeding with Geese," December 30, 2014, *https://www.thepoultrysite.com/articles/weeding-with-geese*
- Metzer Farms, "Using Weeder Geese," *https://www.metzerfarms.com/using-weeder-geese.html*

Chapter 4: Turkeys

- Roy's Farm, "Turkey Breeds—Best 10 Breeds with Pictures," January 15, 2024, *https://www.roysfarm.com/turkey-breeds/*

- Meyer Hatchery, "Heritage Turkeys," *https:// meyerhatchery.com/products/Turkeys-c39791181*
- Hybrid, "Water Consumption Guidelines," *https://www.hybridturkeys.com/en/resources/ commercial-management/feed-and-water/ water-consumption-guidelines/*
- Amy Barkley, "Small Flock Turkey Production, by Penn State Poultry Extension," April 22, 2021, *https://swnydlfc.cce.cornell.edu/submission. php?id=1294&crumb=business|9*
- Kathy McCune, "Is Raising Your Own Meat Turkeys Worth It?" *December 9, 2020, https://familyfarmlivestock.com/ is-raising-your-own-meat-turkeys-worth-it/*
- Poultry Extension, "Space Allowances in Housing for Small and Backyard Poultry Flocks," 2024, *https://poultry.extension.org/articles/getting- started-with-small-and-backyard-poultry/ housing-for-small-and-backyard-poultry-flocks/ space-allowances-in-housing-for-small-and-backyard- poultry-flocks/*

Made in the USA
Monee, IL
03 October 2024

67118698R00216